T0330552

The urban f●rest

in the age of urbanizati●n

Samaneh Sadat Nickayin

RIVER PUBLISHERS SERIES IN CHEMICAL, ENVIRONMENTAL, AND ENERGY ENGINEERING

Series Editors:

MEDANI P. BHANDARI
Akamai University, USA; Summy State University, Ukraine and Atlantic State Legal Foundation, NY, USA

JACEK BINDA
PhD, Rector of the International Affairs, Bielsko-Biala School of Finance and Law, Poland

DURGA D. POUDEL
PhD, University of Louisiana at Lafayette, Louisiana, USA

SCOTT GARNER
JD, MTax, MBA, CPA, Asia Environmental Holdings Group (Asia ENV Group), Asia Environmental Daily, Beijing/Hong Kong, People's Republic of China

HANNA SHVINDINA
Sumy State University, Ukraine
and
ALIREZA BAZARGAN
NVCo and University of Tehran, Iran

The "River Publishers Series in Chemical, Environmental, and Energy Engineering" is a series of comprehensive academic and professional books which focus on Environmental and Energy Engineering subjects. The series focuses on topics ranging from theory to policy and technology to applications.

Books published in the series include research monographs, edited volumes, handbooks and textbooks. The books provide professionals, researchers, educators, and advanced students in the field with an invaluable insight into the latest research and developments.

Topics covered in the series include, but are by no means restricted to the following:

- Energy and Energy Policy
- Chemical Engineering
- Water Management
- Sustainable Development
- Climate Change Mitigation
- Environmental Engineering
- Environmental System Monitoring and Analysis
- Sustainability: Greening the World Economy

For a list of other books in this series, visit www.riverpublishers.com

The Urban Forest in the Age of Urbanisation

Samaneh Sadat Nickayin

Agricultural University of Iceland (AUI), Iceland

River Publishers

Published, sold and distributed by:
River Publishers
Alsbjergvej 10
9260 Gistrup
Denmark

www.riverpublishers.com

ISBN: 978-87-7022-651-6 (Hardback)
         978-87-7022-650-9 (Ebook)

©2021 River Publishers

All rights reserved. No part of this publication may be reproduced, stored in a retrieval system, or transmitted in any form or by any means, mechanical, photocopying, recording or otherwise, without prior written permission of the publishers.

To my grandmother, Sharifeh, who taught me the value of nature

&

To my parents, for all their supports

Contents

Acknowledgements

Writing a book is more challenging than I thought and more rewarding than I could have ever imagined. None of this would have been possible without mosaics of individuals who helped me to finalize it.

I would like to express my special thanks to Fabio Di Carlo, for being a source of inspiration and for all encouragement to pursue my research in landscape studies. I would like to offer my sincere appreciation to Richard Weller, not only for hosting me as a visiting scholar at PennDesign but also for being a great advisor who helped me shape my ideas, thinking deeper.

I am deeply indebted to my loving parents and siblings, without whom I would have never had so many opportunities.

The last word of acknowledgement, I have saved for my grandmother-my coach-life, Sharifeh. Mama Sharifeh, even though you are now far away from us, your garden, your botanist's journey, and your environmental awareness will remain as the background of all my scientific knowledge.

List of Figures

List of Abbreviations

GI   Green infrastructure
TOD   Transit-oriented development
GIS   Green infrastructure stormwater

1

The Urban Forest on the Age of Urbanization: Introduction

1.1 Urban Forest(ry) in the Era of Globalization
Research background

> "The most successful scientist thinks like a poet—wide-ranging, sometimes fantastical—and works like a bookkeeper. It is the latter role that the World sees".

Edward O. Wilson, The Meaning of Human Existence

The book seeks to reflect on the connotation of urban forest(ry),[1] as the central theme of landscape architecture, in line with related emergent holistic theories.

In the era of globalization, when the entire planet is urbanized and planners debate "planetary urbanization", economists discuss "global city", ecologists describe planet's biodiversity hotspots connections, and climate changes warn "global" crisis, it might be reductionist focusing on forests "in" urban and peri-urban edges.

The inception of urban forest(ry) is rooted in the traditional interpretation of the city, where there was a distinct separation between cities and forests. The trees in the city were classified as "urban" forests, and the trees outside of the cities were labelled as wild woodlands. Today, the cities have a planetary scale of influence, and, as geographers, Amin and Thrift (2001) affirm "cities are everywhere, and everything and the footprint of cities are all over". In this new scenario of city, it might be necessary to shift the paradigm of "urban" forest(ry) from traditional "local" perception to a planetary vision.

---

[1] Urban forestry refers to the management of forests in urban area, while urban forest refers to the trees in urbanized area.

1.2 Indian Parable of Blind Men and an Elephant
Problem statement

It has been widely stated that trees in the urbanized area are necessary for our survival—environmentally, economically, and, no less, socially. There are many aspects and approaches towards urban forestry, and experts with different orientation have addressed urban forestry from multiple perspectives and skills. Urban forestry overlaps with managerial discourse (like ecosystem services, biodiversity, and green infrastructure) and "yet no consensus exists on the precise meaning of the terms 'urban forests' and 'urban forestry'" (Krajter Ostoić & Konijnendijk, 2015; Randrup et al., 2005).

It seems like the "famous Indian fable about seven blind men who went to see the elephant. The one who touched the body thought the elephant looked like a wall, another touched the ear and thought the elephant looked like a fan, another touched the tail and thought it looked like a rope, and so on; all were correct in their partial perception of the part but what elephant looked like necessitating a holistic vision".[2]

The holistic approach assembles the potential of other drivers (like economic trajectories, political decisions, and social influences), which force the transformation of the territory and, consequently, has an impact on forestation in the urbanized area. It is not the matter of "what" an urban forest or its function is; it digs into "why" the trees should be planted in a particular place due to specific spatial and contextual pressure.

1.3 Towards Holistic Approach in Urban Forestry
Theoretical background

Theoretically, this book builds upon concepts from theories on environmental ethics, landscape urbanism, and urbanization theories.

The auspices of sages who see the integrity and interconnection of the issues, like Howard Odum, Frederick Law Olmsted, Aldo Starker Leopold, Vandana Shiva, Ian McHarg, Richard Forman, Henri Lefebvre, Neil Brenner, and Richard Weller, become the central concerns that guide this research. It has emphasized the role of "Landscape Architect as urbanist of our age" (Waldheim, 2017) to address the urban forestation according to the upcoming urbanization.

---

[2]Cited by Sajed Kamal (2011). *The Renewable Revolution: How We Can Fight Climate Change, Prevent Energy Wars, Revitalize the Economy and Transition to a Sustainable Future.* London: Routledge; 1 edition p.124.

One of the main criteria that drive the book aims to reverse the concept from "the impacts of forestation on urban area" to "the impacts of the urban area on forestation", not in terms of ecological impacts but in terms of design and spatial configurations of forests. Such choice is rooted in the "potential transformation of context", which is a central issue in landscape architecture.

Each urban area creates a specific "context", which is dynamic and is open to further articulation and could change over time as conditions change. Our planet on a broad scale and our cities on a finer scale are transforming steadily, and without dialectical reasoning between contextual transformation and forestation, any efforts to design the urban forest could result fruitless and fragile in its implementation.

In short, the book tends to highlight the holistic approach of landscape architecture in the urban forest(ry) as an interacting and flexible method, which becomes the driver and "unique way of designing and perceiving territorial transformations, replacing the previous idea of landscape as a balance to urban growth" (Di Carlo, 2015).

1.4 Paradigm Shift Towards the New Edge of Urban Forestry
Research objective and research question

Today, most scientists and experts discuss global rather than local issues. If we accept that all planet is urbanized, and "we share this planet, our home, with millions of species",[3] and "we all breath the same air"[4]; if we admit the global growth of urban population in the next coming years, affirming more soil consumption for urbanization and food production, generating more footprints, and asserting that "climate change does not respect national borders"[5]; if so, in such global urbanized context, defining what is "not" urban forest(ry) and where is its "border line" is meaningless. Most of the school of thoughts have moved towards planetary visions to resolve global environmental issues. The era of holistic approaches necessitates a review of the theory of urban forest(ry).

---

[3]Shiva, V. (2005). *Earth Democracy: Justice, Sustainability and Peace.* Berkeley: North Atlantic Books, p.50.

[4]John F. Kennedy's speech on Peace at the American University, Washington D.C., 10 June 1963.

[5]Ban Ki-moon's speech on Climate change, *Momentum for Change,* at The United Nations, 6 December 2011.

The book moves from dualistic visions of city and wilderness towards holistic theories of cities and natural environments in the modern age through historical studies. Therefore, in line with the aforementioned paradigm shift, the purpose of this book is:

> **To explain why, in the era of "planetary" observations, it is needed to scale up the forest(ry) from the "urban" to "planetary" implication.**

According to this statement, the book aims to find a practical answer that drives the local forestation in urban contexts in line with the global framework. To cover the gap between holistic theories and local implication, the research attempts:

> **To illustrate how an intermediate scale could be defined, based on urbanization "process" and not urban or peri-urban "edges", to address forestation process instead of urban or peri-urban forests.**

In prospect for the future of the planet, with the bourgeoning of next billion people, under the pressure of climate change and *implosion/explosion* of planetary urbanization, the research seeks:

> **To foresee new prospects that forestation in the urbanized area could/should face, in line with upcoming challenges.**

1.5 From Dualism to Holism
Research design

In order to address the central question of the book and fulfil its objective, it has mainly adopted the American school of thoughts. Such choice is embedded in two main reasons. First, the central body of this book was set up on holistic approaches in different fields of study; most of such discussed theories had inception in American academic studies. Second, the opportunity to conduct a research placement in the Landscape Architecture department at the University of Pennsylvania assisted (for the author of this book) to deepen knowledge on emergent global issues. Based on such a theoretical framework, the book comprises five phases.

It begins with a literature study on global "scientific" discourses on urban forestry to delineate principally the scientific approach of foresters, who coined the term "urban forest(ry)".

In the third part of the book, *Dualism: cities versus forests*, the city's theme and its relation to the forests are deepened by addressing philosophical, cultural, and functional bond. In terms of methodology, the first part is based on historical background and its roots in environmental philosophy.

In the fourth part of the book *Holism: Forest in cities/cities in forests*, trends and developments of the new method of "city making", in the era of modernity, are reviewed to outline the continuous updates of disciplines that engage the system of forests and urbanization process. In terms of methodology, environmental ethics become the background of this part, which goes through practical approaches including ecology, urbanization, and landscape theories. This section passes through different scales from local to global, highlighting the necessity of new "planetary" prospects for making forests in urbanized areas. This part attempts to answer the first question of the research.

The parts above theoretical and practical backgrounds are tested in the fifth section of *Urbanization and Forestation*. The case study method is adopted for establishing an intermediate scale between global frameworks and local projects, proving as an answer to the second question of research. To do so, based on the trends of urbanization (developed cities, shrinking cities, and self-constructed cities), the proposition of forestation in each context is discussed. Different scales, locations, and urbanization processes are considered to verify the reliability of the approach in divergent contexts.

The last section of the research defines a paradigm shift in urban forest(ry), in line with emerging global issues of the upcoming century. The relevance mentioned above assumes theoretical and practical references to propose new prospects and practices in the field of urban forest(ry). This part explicated the new hypothesis of forestation in urbanized areas, which could address the rising global challenges.

1.6 From Local to Global Visions
Book structure

This section outlines the reminder of the book. Chapter 3 includes the historical background underlying this book. It provides an overview of the relation between cities and forests. It aims to identify the dualistic approaches that existed between city making and forests management. Furthermore, the chapter highlights the philosophy of Anthropocentrism, as the hidden driver of dualistic choices, which led to the separation of trees from the environment of *civis*. It has been opted to refer to the theory of Max Oelschlaeger on philosophical argumentation, to Robert Pogue Harrison for the mythological studies of trees, and to Michael Williams and Robert W. Miller for practical use of trees, and the historical narration of Henry W. Lawrence defines the cultural framework of this chapter. The chapter closes with some reflections on the inception of urban forestry due to the interdisciplinary collaboration of Olmsted with other experts.

Chapter 4 presents trends and development of holistic theories in the contemporary era, when ecocentrism and environmental ethics tied up the concept of city and nature together, and physical boundaries have vanished. Under the new philosophy of holism, new approaches in ecology, landscape, and urbanization turned up. Landscape urbanism is the central body of this chapter, addressing the forestation issues from regional to global scale. The chapter closes by providing an overview of involved actors and sectors that could guide forestation in urbanized areas. Holistic ecological theories of Howard Odum and Richard Forman and ecological advocacy of Vandana Shiva and McHarg with the advanced hypothesis of planetary urbanization by Henri Lefebvre bring together holistic thinking teleological approach in the era of global linkage.

In Chapter 5, the case study methodology and the results of case studies are presented. This chapter emphasizes the designer's perspective, providing an intermediate scale between global issues and local implication. This chapter includes a wide range of cities with different types of urban tendency to imagine a global city. It can be paraphrased as a collection of "puzzle pieces" that fit together to make a global picture. According to the three types of urban tendency, the chapter discusses the propositions of forestation in each context. Urbanization broadly follows growing trends or shrinking trends; it can be planned by formal or informal initiatives. The discussed categories of urbanizations include developed cities, shrinking cities, and self-constructed cities. In each classification, the hypothesis of forestation is discussed according to the set of circumstances that drive the urbanization process, and the practical methods for forestation are proposed.

In short, this chapter debates how the urbanization tendency defines the structure and spatial composition of forests. It offers a dynamic process of forestation that interacts with the urbanization "process" (and not urbanization "boundaries"). The practical theories from George Hargreaves, James Corner, Michel Corajoud, Gilles Clement, and David Gouverneur intertwine with forestry issues of Claude Guinaudeau, Ingo Kowarik, and Cecil Konijnendijk.

On the grounds of the above considerations, Chapter 6 deliberates on a possible paradigm shift in theoretical contributions and implication for forestation in the urban area in a global framework.

The book concludes by discussing the limitation of the research and exploring avenues for future studies in line with new challenges.

To conclude, I would like to express few words about the title of this book. *The Urban Forests in the Age of Urbanization* is a short and precise title to draw attentions to the planetary dimension of urbanization and its influence on forestation. To express the main proposal of this book, shifting the paradigm of urban forests from "urban edge" to "planetary edge", this chapter has been paraphrased as

The Urban Forest i_on the A_{ED}ge of Urbanization

2

"Exploring Global Scientific Discourses on Urban Forestry": Literature Review

"Exploring global scientific discourses on urban forestry"[1]

> "If I have seen a little farther than others, it is because I have stood on the shoulders of giants".

Isaac Newton

At the University of Toronto, seeking an appropriate title of a graduate student's thesis, in 1965, Prof. Erik Jorgensen proposed the unusual juxtaposition of "urban" and "forestry" (Gerhold, 2007). The term was coined as a branch of forestry that deals with "cultivation and management of trees for their present and potential contributions to the physiological, sociological, and economic well-being of urban society" (Jorgensen, 1970). Furthermore, Prof. Jorgensen believed that "urban forestry not only dealt with city trees or with single tree management but also with tree management in the entire area influenced by and utilized by the urban population" (Jorgensen, 1970; Konijnendijk et al., 2006).

Later, the plantation of trees in urban areas has introduced other concepts. For instance, Guinaudeau (1987) in France introduced the concept of *Préverdissement*. This concept is based on the planting *prior to* urbanization development to arrange the plants with planned structures or construction (plant today, built tomorrow). It is considered a combination of design and protection to reach the desired size of plants and guarantee the aesthetic quality of spaces (Craul, 1999). In the middle of the 1990s, the term green infrastructure has emerged based on urban forestry.

---

[1]Refers to the title of Krajter Ostoić, S., Konijnendijk, C.C. (2015). Exploring global scientific discourses on urban forestry. *Urban Forestry & Urban Greening* 14 (2015) p. 129–138.

Beyond the inception of urban forestry and its evolution, it has been widely stated that forests in urbanized areas can address a wide range of concerns, such as environmental, economic, and social issues. Environmental issues deal with air pollution, and Carbone sequestration, wildlife habitat, water quality, and climate change; the economic concerns are related to food and timber production; social issues refer to human health and individual or collective well-being other sets of semantic and intangible values. In this sense, urban forests are the central concern of ecosystem services, providing complex services to humankind.

According to the main challenges in each context, such as the type of urban expansion, history, social, and economic trends, urban forests aim at specific targets.

The central issues that will come up in this book are based on an understanding of the most predominant urban forestry approaches in the world. In order to assemble such a global framework of theories, this section reviews the primary literature of each continent.

2.1 From Urban Forest to Woodland
North America and Europe

The leading research in urban forestry has been conducted by North American and followed by European scientists[2] (Krajter Ostoić and Konijnendijk, 2015), encompassing multiple benefits of urban forestry. Europe dominantly drew attention to woodlands, while studies in North America focused on trees and "urban forest"; such divergence is rooted in the historical background of urban forestry in Europe and North America (Konijnendijk, 2006). In Europe, the aesthetic and recreational benefits were considered central issues of trees plantation (like parks, gardens, etc.), while in North America, environmental services have been prioritized and the inception of urban forestry rooted in environmental issues.

North America mainly emphasizes the quantification of urban forests benefits[3] (Rowntree and Nowak, 1991; Dwyer et al., 1992; Nowak et al., 2010) and connectivity of green infrastructure (Benedict and McMahon,

---

[2]It is stated that in the period of 1988-2014, 317 publications have been published by American and Canadian researcher and 104 publications by European scientists. For details of countries and numbers, see Krajter Ostoić and Konijnendijk (2015).

[3]For instance, i-Tree program was developed by United States Forest Service to quantify the ecosystem services provided by trees.

2006; Benedict and McMahon, 2001). Recently, research has been developed on urban forest vulnerability assessment, and its adaptation to climate change and the arduous growing conditions of urban forests open new debates among experts (Brandt et al., 2016; Calfapietra et al., 2015). Nonetheless, it is a noticeable lack of discussion on climate change mitigation and adaptation (Krajter Ostoić and Konijnendijk, 2015). In Europe, the productivity of urban forests[4] (Tyrväinen, 2001; Rydberg and Flack, 2000; Sæbø et al., 2003), social issues, and recreational use (Arnberger, 2006) have been chiefly argued.

Since the exact domain of urban forestry is not well-defined, the managerial discourse like ecosystem services, biodiversity conservation, and green infrastructure overlap with urban forestry (Krajter Ostoić and Konijnendijk, 2015).

Recently, urban forestry has been adopted in Nature-Based Solution[5] (NBS) projects to enhance climate and water resilience in urban areas. NBS proposes interdisciplinary techniques, from macroscale (global) to mesoscale (national) and microscale (local), to address the gap between science, policy, and society (Raymond et al., 2017). Even though NBS is still in its infancy, the list of knowledge provided by practitioners offer a combination of different existing approaches as a framework to guarantee the achievement of the expected objectives and solves the targeted problem under different uncertain future situations.

2.2 Pollution
Asia

Most of the largest cities in the world are situated in Asia,[6] hosting 60% of the world's population,[7] and it has only 17% of global forests

---

[4]The productivity can include recreational profits, silvicultural, or timber production (timber production is often of lower importance).

[5]"Nature-Based Solutions (NBS) are solutions to societal challenges that are inspired and supported by nature, which are cost-effective, provide simultaneous environmental, social and economic benefits, and help build resilience" (Raymond et al., 2017); they provide benefits for both biodiversity and human well-being (Cohen-Shacham et al., 2016). See EKLIPSE project funded by European commission under the Horizon 2020. Available at: http://www.eklipse-mechanism.eu/about_eklipse

[6]We can mention Shanghai, Beijing, Karachi, Istanbul, Dhaka, Tokyo, Manila, Tianjin, Mumbai, Shenzhen, Guangzhou, Delhi, Wuhan, Lahore, Seoul, and Chengdu are among the first 20 largest cities in the world. See http://www.citymayors.com/statistics/largest-cities-mayors-1.html

[7]See http://www.worldometers.info/world-population/asia-population/

(WWF Report, 2012). With the expansion of the cities, the gap between ecological footprint and biocapacity is widening.[8]

What is the main challenge that forestation should face in Asian cities? Asia accommodates not only the largest cities but also the most polluted cities[9] in the world. The afforestation in an urbanized area mainly attempts to compensate for the pollution. For instance, China is a pivotal contributor to CO_2 emission (Chen et al., 2007; Global Carbon Project, 2008); so it became one of the most diligent countries to enhance carbon sequestration by forestation. In fact, most of the research in China debates the impacts of urban forests on offsetting carbon emissions in different cities around China (Yang et al., 2005; Liu and Li, 2012; Zhao et al., 2010; Zhao et al., 2010; Chen, 2015). Due to the lack of techniques for mixed plantation and to facilitate operations in silviculture, the urban forests in China are mainly monocultures (Liu et al., 2004). Similar researches regarding carbon sequestration have been conducted in Korea (Jo, 2002), Iran (Faryadi et al., 2009), Pakistan (Kazi, 2016; Akbar et al., 2014), India (Avni and Chaudhry, 2016), Malaysia, and the Philippines (Kuchelmeister, 1998). Urban forestry in Asia concerns with other issues like poverty alleviation conducted in Bangladesh (Uddin, 2006) and human well-being (*Shinrin-yoko* or forest bathing) in Japan (Park et al., 2010).

Nonetheless, urban forestry remains a neglected area of research in many Asian countries, and "multipurpose urban forestry is even more in its infancy" (Kuchelmeister, 1998). To wrap up, by virtue of environmental health hazards, carbon sequestration is the main issue in many Asian countries.

2.3 From Woodfuel to Luxury Gradient and Desertification
Africa

In Africa, more than 80% of all domestic fuel is provided by wood; the main study proposes the complex relation between towns and forests, focusing on

---

[8]According to Global Footprint Network, the ecological footprint is an accounting framework to measure the amount of biologically productive land and sea area that humanity needs to produce the resources it consumes, provide room for its infrastructure, and absorb its waste. The biocapacity is the environment's ability to restore the resources; in Asia, there is a shortfall of 0.8 global hectares per person (WWF Report, 2012).

[9]According to WHO (World Health Organization, 2016) the most polluted cities in the world are Zabol (Iran); Gwalior, Allahabad, Patna, and Raipur (India); Riyadh and Al Jabail (Saudi Arabia); and Xingtai and Baoding (China).

the woodfuel issue [10] (FAO, 2012; Drigo and Salbitano, 2008). For instance, the Maamore forest in Morocco, the gallery forests of the Batékés highlands in the Republic of the Congo (Schur et al., 2011), the eucalyptus plantations of the Malagasy highlands in Madagascar, the dry forests and agroforestry systems of Mali, and the riparian forests of South Africa, all play essential roles in supplying Africa's urban areas with woodfuel (FAO, 2016).

Another critical issue in Africa is desertification. *The Great Green Wall of the Sahara and the Sahel Initiative* aims to stop the southern advance of Sahara.[11] Such a broad-scale approach has also been adopted in a fine-scale, like the city of Ouarzazate in Morocco, where the plantation of Green Belt, called "the door of the desert", has been aimed to stop the desertification and land degradation, protecting cities from winds and dust of desert (UNEP, 2015).

Being the second most populated (16.3% of the world's population) and the poorest continent globally, food and nutrition security turn into urban forestry issues. To tackle poverty and inequality, the report of Food and Agriculture Organization (FAO, 2016) proposed a novel method of urban forestry called "*luxury*" *gradient*. Such a proposal is based on the choice of species, "ranging from purely aesthetic considerations" (for wealthier neighbourhood) to the more "pragmatic provision of goods and services" (for more impoverished neighbourhood). For instance, there is an unequal distribution of trees between wealthy northern and poorer southern districts of Johannesburg (Schäffler and Swilling, 2013). By implementing the "luxury gradient" approach in the poor district of Johannesburg, not only can trees contribute to the mitigation of poverty, but they can also enhance the tree canopy in the southern district. Furthermore, in southern Africa, traditional medicine as an alternative healing practice is common among communities (Geldenhuys, 2007); so access to the indigenous trees in urban forest can support the traditional curative method.

To wrap up, in the African context, desertification, alleviation of poverty, and woodfuel are the main concerns in urban forestry.

---

[10]See also "WISDOM for Cities" (Woodfuel Integrated Supply/Demand Overview Mapping), a tool for mapping sustainable resource and potential area for woodfuel in east Africa and Southeast Asia (Drigo and Salbitano, 2008).

[11]It is stated that the green wall has been proposed for the first time by Richard St. Barbe Baker (1889-1982) in 1952 calling for a 30-mile-wide "Green Front". See http://wilmetteinst itute.org/the-man-of-the-trees-and-the-great-green-wall-a-bahais-environmental-legacy-fo r-the-ages/

2.4 Economic Extraction
Latin America

Half of the tropical forests and one-fourth of the world's total forests lie in the Latin America region, providing vital global and national environmental services.[12] Due to the economic extraction and urbanization near/in natural forests (like Favelas), a dramatic change has occurred in land use and natural resource. As the vicinity and the influence of urbanized areas on natural forests cannot be ignored, the priority to the "peri-urban", rather than "urban" forestry in Latin America is mainly stated (FAO, 2016).

The main effort in Latin American contexts is concentrated on the "conservation" of natural resources from urbanization effects, providing economic opportunities for the urban dwellers.[13] For instance, the National Park of Tijuca in Rio de Janeiro is considered the most extensive urban forest globally, which has been frequently damaged by fire;[14] to decrease the fire frequency, it has been planted less-flammable plants in the forests (Silva Matos et al., 2002). Other issues like inequality distribution of trees and ecosystem services in Latin American cities (which lead to environmental injustice) are discussed by experts (Escobedo et al., 2015).

Apart from a few "urban" forestry cases in Latin America,[15] the leading research aims at peri-urban forestry and conservation of natural forests, providing job opportunities for the young generation. For instance, in Niterói (one of the municipalities of Rio de Janeiro), it launched the social integration program of *"Nem-Nem": Nem estuddam, nem trabalham*[16] to provide job opportunities and income for the young unemployed generation, who contribute to the recovering of environment.

To conclude, each context necessitates a specific target for urban forestry; in the planetary vision of forestation, it is essential to consider the most predominant drivers of each context.

---

[12] See http://www.fao.org/docrep/006/J2459E/j2459e12.htm

[13] Two proposals for urban forestry in Latin America are debated in the chapter of case study: Chamanga and Cartagena.

[14] It is reported that the main cause of fire included rubbish burning, hot air balloons, and intentional and religious practices (Silva Matos et al., 2002).

[15] For instance, São Paolo's public–private partnerships for tree-planting; by planting trees, the company receives an advertising spot (Zulauf, 1996; Coleman et al., 2013) or the city of Curitiba (Brazil) has adopted the BioCity Program to integrated the biodiversity in urban planning (FAO, 2016; UNEP, 2008) or reduction of air pollution in Santiago by urban forestry (Escobedo et al., 2008)

[16] See http://www.fao.org/forestry/urbanforestry/86956/en/Presentation n.32 by Axel Grael: Programa NITER.

References

Akbar, K.F., Ashraf, I., Shakoor, S. (2014). Analysis of urban forest structure, distribution and amenity value: A case study of Sahiwal. *The Journal of animal and plant Sciences* 24(6), p.

Arnberger, A. (2006). Recreation use of urban forests: An inter-area comparison. *Urban Forestry & Urban Greening* 4(3-4). p. 135-144.

Avni, A., Chaudhry, P. (2016). Urban Vegetation and Air Pollution Mitigation: Some Issues from India. *Chinese Journal of Urban and Environmental Studies* 4(1).

Benedict, M.A., McMahon, E.T. (2006). *Green infrastructure: Linking landscape and community*. 2nd edition. Washington D.C.: Island Press.

Benedict, M.A., McMahon, E.T. (2001). Green infrastructure: Smart conservation for the 21st century. *Sprawl Watch Clearinghouse Monograph Series*.

Brandt, L., Lewis, A.D., Fahey, R., Scott, L., Darling, l., Swanston, C. (2016). A framework for adapting urban forests to climate change. *Environmental Science and Policy* 66. p. 393-402.

Calfapietra, C., J. Peñuelas, and T. Niinemets. (2015). Urban plant physiology: Adaptation-mitigation strategies under permanent stress. *Trends Plant Sci*. 20:72–75.

Chen, J. M., Thomas, S. C., Yin, Y., Maclaren, V., Liu, J., Pan, J., Liu, G., Tian, Q., Zhu, Q., Pan, J.-J., Shi, X., Xue, J., & Kang, E. (2007). Enhancing forest carbon sequestration in China: Toward an integration of scientific and socio-economic perspectives. *Journal of Environmental Management*, 85, p. 515–523.

Chen, W.Y. (2015). The role of urban green infrastructure in offsetting carbon emissions in 35 major Chinese cities: A nationwide estimate. *Cities* 44, p.112–120.

Cohen-Shacham, E., Walters, G., Janzen, C., Maginnis, S. (2016). *Nature-based Solutions to address global societal challenges*. IUCN, Gland: Switzerland.

Coleman, G., Kontesi, A., Li, X., Masliah, A., Renwick, D., Torà, L. & Vargas, L. (2013). *Building successful public private partnerships in São Paulo's transportation sector*. Columbia University School of International and Public Affairs. Available at: www.usp.br/iri/images/anexos/pesq_Joa oCV_Columbia_University-buildingsuccessful-ppps-in-sao-paulo.pdf). [Accessed 20 September 2017]

Craul, P.J. (1999). *Urban Soils: Applications and Practices.* New York: Wiley; 1 edition, p. 243-244.

Drigo, R., Salbitano, F. (2008). WISDOM for Cities: analysis of wood energy and urbanization using WISDOM methodology. Rome. FAO. Available at: http://www.fao.org/fileadmin/templates/FCIT/PDF/WISDOM.pdf [Accessed 19 September 2017]

Dwyer, J.F., McPherson, E.G., Schroeder, H.W., Rowntree, R.A. (1992). Assessing the benefits and costs of the urban Forest. *Journal of Arboriculture* 18 (5). p. 227–234.

Escobedo, F. J., Clerici, N., Staudhammer, C. L., & Corzo, G. T. (2015). Socio-ecological dynamics and inequality in Bogotá, Colombia's public urban forests and their ecosystem services. *Urban Forestry and Urban Greening*, 1040-1053.

Escobedo, F.J., Wagner, J.E., Nowak, D.J., De la Maza, C.L., Rodriguez M., Crane, D.E. (2008). Analyzing the cost effectiveness of Santiago, Chile's policy of using urban forests to improve air quality. *Journal of Environmental Management* 86(1), p. 148-157.

FAO. (2016). *Guidelines on urban and per-urban forestry.* By Salbitano F., Borelli, S., Conigliaro, M., Chen, Y. FAO Forestry Paper No.178. Rome.

FAO. (2012). *Urban and peri-urban forestry in Africa: the outlook for woodfuel, Urban and peri-urban forestry working paper n°4.* Rome.

Faryadi, Sh., Taheri, Sh. (2009). Interconnections of Urban Green Spaces and Environmental Quality of Tehran. *International Journal of Environmental Research* 3(2), p. 199-208.

Geldenhuys, C.J. (2007). Restoring natural forests to make medicinal bark harvesting sustainable in South Africa. In: Aronson, J., Milton, S.J., Blignaut, J.N. (Eds.). *Restoring Natural Capital: Science, Business, and Practice.* Island Press, pp. 170–178 (Chapter 20).

Gerhold H.D. (2007). Origins of Urban Forestry. In: Kuser J.E. (eds) *Urban and Community Forestry in the Northeast.* Dordrecht: Springer.

Global Carbon Project (2008). *The global carbon budget 2007.*

Jo, H. (2002). Impact of urban greenspace on offsetting carbon emissions for middle Korea. *Journal of Environmental Management* 64(2), p.115-126.

Jorgensen, E. (1970). Urban forestry in Canada, *Proc. 46th Intl. Shade Tree Conf.*, 43a–51a.

Kazi, M. (2016). Pakistan's first urban Forest makes way in concrete jungle. [online] *The Express Tribune.* Available at: https://tribune.com.pk/story/1145932/breathing-space-pakistans-first-urban-forest-makes-way-concrete-jungle/ [Accessed 16 September 2017]

Konijnendijk, C.C., Ricard, R.M., Kenney, A., Randrup, T.B. (2006). Defining urban forestry – a comparative perspective of North America and Europe. *Urban Forestry and Urban Greening* 4, p. 93–103.

Krajter Ostoić, S., Konijnendijk, C.C. (2015). Exploring global scientific discourses on urban forestry. *Urban Forestry & Urban Greening* 14 (2015) p. 129–138.

Kuchelmeister, G. (1998). Urban forestry in the Asia-Pacific Region – status and prospects. Asia-Pacific Forestry Sector Outlook Study Working Paper Series No. 44. Rome, FAO. Available at: www.fao.org/forestry/FON/F ONS/outlook/Asia/APFSOS/44/Apfsos44.htm [Accessed 15 September 2017]

Liu, C., & Li, X. (2012). Carbon storage and sequestration by urban forests in Shenyang, China. *Urban Forestry & Urban Greening*, 11, p. 121–128.

Liu, C., Shen, X., Zhou, P., Che, Sh., Zhang, Y., Shen, G. (2004). Urban Forestry in china: Status and Prospects.*UA-Magazine*. p. 15-17.

Nowak, D.J.; Stein, S.M., Randler, P.B., Greenfield, E.J. Comas, S.J., Carr, M.A., Alig, R.J. (2010). Sustaining America's Urban Trees and Forests. Available at: http://unri.org/ECO%20697U%20S14/nrs-62_sustaining_a mericas_urban%20copy.pdf [Accessed 19 September 2017]

Park, J., Tsunetsugu, Y., Kagawa, T., Miyazaki, Y. (2010). The physical effects of *Shinrin-yoko* (taking in the forest atmosphere or forest bathing): evidence from field experiments in 24 forests across Japan. *Springer, Environmental Health and Preventive Medicine* 15(1). p. 27–37.

Raymond, C.M., Berry, P., Breil, M., Nita, M.R., Kabisch, N., de Bel, M., Enzi, V., Frantzeskaki, N., Geneletti, D., Cardinaletti, M., Lovinger, L., Basnou, C., Monteiro, A., Robrecht, H., Sgrigna, G., Munari, L. and Calfapietra, C. (2017). *An Impact Evaluation Framework to Support Planning and Evaluation of Nature-based Solutions Projects*. Report prepared by the EKLIPSE Expert Working Group on Nature-based Solutions to Promote Climate Resilience in Urban Areas. Centre for Ecology & Hydrology, Wallingford: United Kingdom

Rydberg, D., Flack, J. (2000). Urban forestry in Sweden from a silvicultural perspective: a review. *Landscape and urban planning* 47 (1-2). p. 1-18.

Rowntree, R.A., Nowak, D.J. (1991). Quantifying the role of urban forests in removing atmospheric carbon dioxide. *Journal of Arboriculture* 17 (10). p. 269–275.

Sæbø, A., Thorarrin, B. Randrup, B.T. (2003). Selection for trees for urban forestry in the Nordic countries. *Urban Forestry and Urban Greening* 2. p. 101–114.

Schäffler, A., Swilling, M. (2013). Valuing green infrastructure in an urban environment under pressure-The Johannesburg case. *Ecological Economics* 86, p. 246–257.

Schur, J., Ingram V., Marien, J'N., Nasi, R., Dubiez, E. (2011). *Woodfuel for urban centers in the Democratic Republic of Congo.* Brief No.7. Bogor, Indonesia, Center for International Forestry Research.

Silva Matos, D.M., Santos, C.J.F., Chevalier, D.R. (2002). Fire and restoration of the largest urban Forest of the World in Rio de Janeiro City, Brazil. *Urban Ecosystems*, 6, p.151–161.

Tyrväinen, L. (2001). Economic valuation of urban forest benefits in Finland. *Journal of environmental management* 62(1). p. 75-92

Uddin, M.N. (2006). The relationship between urban forestry and poverty alleviation: Dhaka as a case study. Available at: http://www.fao.org/upload s/media/The_relationship_between_Urban_forestry_and_poverty_allevia tion_Dhaka_case_study.pdf [Accessed 10 September 2017]

UNEP. (2015). Moroccan city defies desertification by harnessing solar power and treated wastewater. UNEP News Center. United Nations Environment Program (UNEP). Available at: http://www.unep.org/newscentre/morocc an-city-defies-desertification-harnessing-solar-power-and-treated-waste water [Accessed 19 September 2017]

UNEP.(2008). *City of Curitiba, Brazil BioCity Programme: mainstreaming biodiversity.* Brochure. United Nations Environment Programme (UNEP) Available at: www.unep.org/urban_environment/PDFs/Curitiba_Final.P DF. [Accessed 18 September 2017]

WWF Report. (2012). *Footprint and investment in natural capital in Asia and the pacific.* Manila: Printing Services ADB.

Yang, J., McBride, J., Zhou, J. and Sun, Z. (2005). The urban Forest in Beijing and its role in air pollution reduction. *Urban Forestry & Urban Greening* 3, p. 65-78.

Zhao, M., Kong, Z., Escobedo, F. J., & Gao, J. (2010). Impacts of urban forests on offsetting carbon emissions from industrial energy use in Hangzhou, China. *Journal of Environmental Management*, 91, p. 807–813.

Zhao, J., Ouyang, Z., Xu, W., Zhang, H., & Meng, X. (2010). Sampling adequacy estimation for plant species composition by accumulation curves – A case study of urban vegetation in Beijing, China. *Landscape and Urban Planning*, 95, p. 113–121.

Zulauf, W. (1996). Legal, institutional and operation structure of urban green-area systems. *Paper presented at Urban Greening Seminar*, Mexico City, Mexico, p. 2–4.

3

Dualism: Cities Versus Forests

Cities versus forests

> "Much of unconscious life of the individual is rooted in interaction
> with otherness that goes beyond our own kind, interacting with
> it very early in personal growth, not as an alternative to human
> socialization, but as an adjunct to it [. . .] Identity formation grows
> from the subjective separation of self from non-self, living from
> non-living, human from nonhuman, and proceeds in a speech to
> employ plant and animal taxonomy as a means of conceptual
> thought and as a model of relatedness".[1]

Paul Shepard

During the history of civilization, the human species acquire triumph over
the inhospitable environment, and such human supremacy leads to the
idea of dualism between "civilized" and "wilderness" (Oelschlaeger, 1991).
Civilized World or cities became the natural habitat of *civis*,[2] and wilderness[3]
became the resource for *civis*.

For many centuries, the dualism between human and nature had echoes in
the assertion of many intellectuals like as Francis Bacon who, in 1620, argued
that "the world is made for man, not man for the world" or Sigmund Freud
(1927), who expressed that "the principal task of civilization, its actual raison
d'etre, is to defend us against nature".

This distinction between the civilized World and wilderness places the
human at the centre of the universe, the human becomes the most powerful
entity in the universe, and the reality is assessed through an exclusively

---

[1]Shepard. P. (1982). *Nature and Madness*. San Francisco: Sierra Club Books, p.125.

[2]Latin word of *civis* – means city-dweller and the word of 'civilization' refers to it.

[3]Wood may have a common root with wild, and the word *savage* is derived from *sylva* _ "a
wood" (Williams, 2006).

human perspective, what Dave Forman[4] (1991) called Anthropocentrism.[5] Anthropocentrism is dualist in origin: humanity controls nature for its benefits or purpose. Anthropocentrism has been the main reason for the extraction and alternation of many natural resources like forests.

Based on historical background, this chapter argues that forestation in and around cities has reflected the concept of Anthropocentrism from early civilization to the present day.

According to the historical circumstance, humans' character has evolved, and humans achieved new knowledge over nature. Planting trees has reflected the evolving character of human existence from early civilization to the present day. To illustrate, in prehistoric societies, the forests become invested with sacredness, and later with the evolution of agriculture, the forests were perceived as a resource, and with the birth of gardens, trees were planted for embellishment.

All these historical sequences are studied to highlight how Anthropocentrism manipulate the presence of trees in our cities.

3.1 Sacredness

In many cultures, forests and trees have represented "special spiritual significance to people and communities" (Oviedo and Jeanrenaud, 2007) and are associated with secrecy and initiation rites. For many reasons, trees and forests have been considered sacred; in some cultures, the trees were correlated to birth, marriage, death, fertility,[6] tradition, or religion, and in others, they have been the residence of spirits or have been identified as national monuments[7] (Barrow, 2010).

Sacredness is rooted in religion. Such a religious view of trees is deeply rooted in the image that human beings have of self and nature (Nasr,

---

[4]Dave Forman is the co-founder of Earth First, which is a radical environmental advocacy, founded in 1980 with the goal to draw attention upon environmental conservation.

[5]It is also known as human exceptionalism (Catton and Dunlap, 1978).

[6]Crews highlights that "Because of the shape of trees – a central trunk with branches like arms and fingers, bark like skin – trees lend themselves to identification with the human form, and have frequently been endowed symbolically with anthropomorphic characteristics, leading to a link with fertility symbols in some cultures". See: http://www.fao.org/3/y9882e/y9882e08.htm.

[7]"Certain trees are national monuments and appear on the national flags, for example in Lebanon it is the Cedar of Lebanon (*Cedrus libani*), Canada the Maple (*Acer saccharum*) and for Chile the Monkey Puzzle tree (*Araucaria araucana*)" (Barrow, 2010, p. 44).

1998). Such imagination of self can be based on Anthropocentrism or ecocentrism.[8]

Eastern religions (Buddhism, Hinduism, Shintoism, and Taoism) and the cosmovision of most indigenous and traditional peoples correlate to nature in more ecocentric terms, whereas Abrahamic religions (Islam, Judaism, and Christianity) perceive nature in more anthropocentric terms (Harmon and Putney, 2003). Abrahamic religions, in particular Judeo-Christian, have ambivalent and antithetical messages. On the one hand, they emphasize stewardship towards nature; on the other hand, they accentuate domination over nature (Williams, 2006). Sacred groves are examples of stewardship, while deforestations[9] are examples of domination over the forests (Williams, 2006; Barrow, 2010).

In his book *Forest: The Shadow of Civilization*, Harrison (1992) mentions that in the early Mediterranean civilizations, early deforestation was "religious actions" to permit primitive people to see the sky better to read the divine signs sent down to humans from an abstract "above". In Greek and early Christians, the Earth was "an abode designed for humankind" and "*man*kind was created to rule over God's earthly creation" (Oelschlaeger, 1991), where nature must be conquered through pain and suffering; this is what Thomas Aquinas described as "punishment of original sin".

The dominant idea, and clearing the forest was due to the idea that the forest was associated with pagan gods and devils[10] and make a clearing in forest was a means of purifying the faith, and it became a sanctuary against evil (Williams, 2006). Forests were a palace to draw nearer to God; humans were blessed, and they assisted both God and themselves in the improvement of an earthly home.[11]

In other visions that focus on stewardship, sacred groves as the gods' abode have been venerated and protected for millennia. Even though scared groves exist in most of the world, they are more associated with India than any other country, where the concept of sacred groves predates the Vedic age (Ramakrishnan, 1996). Sacred groves are also connected to Buddhism in

---

[8]Refers to environmental ethics that studies the moral relationship of human beings to the environment and its non-human contents.

[9]"The link between Western Christian piety and land reclamation was a leitmotif running through medieval clearing" (Williams, 2006, p. 96).

[10]In popular folk cultures, the forests were the residents of demons, monsters, witches, and werewolves.

[11]In Christian view of thing, the earth was merely "a sojourner's way station".

China and Japan, where the trees were planted near temples. Persian gardens were equivalent to sacred groves, where fundamental elements of the universe come together (Grimal, 1974).

The importance of sacred groves could also be attested in many sacred texts; the Date palm is mentioned in the Bible, Qur'an, and Thalmud (Hareuveni, 1980; Waisel and Alon, 1980). One hadith[12] states: "When doomsday comes if someone has a palm shoot in his hand, he should plant it". It implies that, even when all hope is lost, such planting is good in itself (Khan, 1999).

To wrap up, under the idea of sacredness, the human had applied two different approaches to control the forests. One led to deforestation (Williams, 2006), and the other attends the stewardship of forests (Barrow, 2010). In such Anthropocentrism, woods and forests were located outside of the habitat of *civis* because forests provoked secrecy and fear, and as humans were not able to explain them, they associated the forests with the place for divinities.[13] Under the idea of sacredness, humans set up boundaries between self and "others".

3.2 Resourcism

> "I marvel at how our God has given so many uses for wood for all men in the whole wide World: building timber, firewood, joiner's, cartwright's and shipbuilder's wood, wood for rooms, wood for wheelbarrows, paddles, gutters, barrels. In short, wood is one of the greatest and most necessary things in the World, which people need and cannot do without".

Martin Luther [14]

Forests have been considered as resources for many purposes. In Anthropocentrism, a "nonhuman world, which is all external to man and his structures", are considered "as raw material [...] to the human purpose" (Livingston, 1985). Neil Evernden (1984) says that "Resourcism is a

---

[12]Saying of the Prophet Mohammed.

[13]For instance, in the beliefs of prehistoric societies, the trees struck by lightning and consumed in the resulting fire, have associated with divinities who dwelled the heavens as well as the earth (Brosse, 1989).

[14]Martin Luther *Table Talk* of 30 August 1532, cited in *Wood, A history* by Joachim Radkau (2011).

modern religious which casts all of creation into categories of utility" and Oelschlaeger (1991) explains it as "transformation of modern people from Homo religious to *Homo economicus*".[15]

Although resourcism is a contemporary idea, its inception returns to the Neolithic revolution and agricultural developments. Resourcism has led to the deforestation and cultivation of domesticated trees.[16] Since the first permanent settlement and domestication of plants, the trees were cultivated for food, medicine, and shadow in the cities. Drawings and descriptions of pre-Columbian America suggest that many native American tribes developed extensive agricultural communities, including extensive gardens with planted trees (Miller et al., 2015). In China, Kublai khan necessitated tree planting along all public roads in and around Beijing for shade (Profuse, 1992). The early Egyptians described the journey of trees with balls of soil to be transplanted in cities more than 4000 years ago (Chadwick, 1971).

New necessities regarding the management of trees as resources have emerged. The history of professions has followed the path of resourcism and new professions such as forester, arboriculture, gardener, plant hunters, and botanist have appeared.[17] Arboriculture, as we know it today, has risen between 1400 and 1800; James Lyte's book *Doden* (1578) used the term of the arborist, and William Lawson's *A New Orchard and Garden* (1597) discussed the methods of planting, pruning, fertilizing, and wound treatment (Gerhold and Frank, 2002).

Resourcism is closely linked to the great voyages of discovery of plant hunters to find new sources for food, seeds, and medicine worldwide.[18]

Resourcism, in its anthropocentric vision, lacked any awareness of global responsibility and "heroic exploits" of plant hunters traded endangered

---

[15]See also Caruso, S. (2012). *Homo oeconomicus. Paradigma, critiche, revisioni.* Firenze: Firenze University Press.

[16]The forests burned down to improve hunting ground, increase yield of the gathered fruits, harvest shoots from coppiced woods for better building, and weaving material to gain open land for farming and for energy production. Animal husbandry and the need for fodder and grazing of livestock and pollarding have also influenced on forests (Williams, 2006).

[17]See also the concept of *Homo reciprocans*, which refers to the human, who improves the environment and also achieves his needs and wants.

[18]See also Bobby J. Ward. (2004). *The Plant Hunter's Garden. The New Explorers and Their Discoveries.* Portland: Timberpress; Tyler Whittel, M. (1997). *The Plant Hunters: Tales of the Botanist-Explorers Who Enriched Our Gardens (Horticulture Garden Classic).* Guilford: Lyons Press; 1st edition (translated by B. Bini in Italian) and in Italian version: Zanazzi, L. (2013). *Uomini e piante. Le passioni dei collezionisti del verde.* Roma: Deriveapprodi.

species shamelessly. Stefen Schneckenburger (2001) in his writing about plant collector Albert Purpuse (1851-1941) expressed that:

> "It is doubtless a good thing that the age is now past in which plant hunters and botanists collected in great quantities. A range of laws and regulations relating to nature conservancy, the Washington Convention on International Trade in Endangered species (CITES) and biodiversity conventions are, on the one hand, providing urgently needed protection for natural environments and individual species and, on the other, protecting the interests of countries in the use of their own natural resources. Just as it would be a mistake to condemn past collectors from present day perspective, we would also be in error if we were to cite their behaviour as a standard for the twenty-first century at a time when nature is being wantonly sold off on a worldwide scale. Rather than regarding these plant hunters as a paradigm, we should see them as children of their day and age, when they collected the bounties of a seemingly inexhaustible natural environment, at times under extremely harsh conditions and economic constraints".

From the present-day perspective, the resourcism is mainly considered cruel against the environment, albeit it had played an essential role in the history of urban tree planting. Under resourcism, the first trees were planted within the city's wall, in the private gardens of ruling classes for utilitarian purposes (fruit, woods, etc.) rather than ornament (Lawrence, 2008).

3.3 Embellishment[19]

> "Beauty will save the World".
> Fëdor Dostoevskij

> "Wildness is the preservation of the World".
> Henry David Thoreau

Since the early garden, trees were planted to provide a perceptual experience of beauty and pleasure. The sense of beauty was defined as the result of the rational order of planting, which provides harmony and symmetry with aesthetic values. If we consider that "chaos is the law of nature [and] order is the dream of man"—as Henry Adam said—the idea of embellishment is essentially anthropocentric. *Homo aestheticus*[20] imposed his idea of beauty in nature, shaping trees, designing gardens as a geometrical abstraction of nature.

In Europe, from the 16th to the 17th century, there was a steady increase in the number of *Homo aestheticus* in research of beauty in the places planted by trees. Lawrence (2008) explains that during this period, nationally distinct ways of using trees have risen: Renaissance gardens developed among the landscape of Rome, Paris, and The Hague, and trees in Renaissance and Baroque garden were kept small, only the *bosco* outside of the walls could have their full size; French brought further use of rows of trees in the form of the *allées*, *mails*, *cours*, and *Boulevards*; Dutch created the tree-lined canal and *Lange Voorhout*.[21] Such an aesthetic approach opened new criticism. For instance, Camillo Sitte (1965), who was the most vocal opponent of the Haussmann-Stübben, opposed that:

> "[...] All tree-lined streets are tedious, but no cities can do without them [and boulevards] thousands of trees would have been more usefully planted in two or three new parks [...] The motif of the *Alleé* is in itself a burning indictment of our taste [...]. One literally gets a headache from such oppressive boredom.

---

[19]Embellishment derives from the French term *embellissement* which expresses an idea that is more complex than simple niceties. It refers also to the quality of urban environment; for instance, in the essay of Voltaire (1749) *"Des Embellissemens de Paris"*, it is mentioned the need for public buildings, bridge, market, and fountains.

[20] Title of the book Ellen Dissanayake (1992) *Homo Aestheticus: Where Art Comes from and Why?*

[21]The front gardens in The Hague that unified to create an L-shaped broad avenue.

> And this is the major "art form" of our city planners of the
> geometric persuasion [...] faults of modern arrangements derive
> solely from the fact that all tree-lined alleys have been planned
> on the drafting board according to the principles of symmetry,
> without consideration for the well-being of the vegetation, light
> and sun, or the consequent effect on cityscape and traffic".

Albeit, embellishment expressed a new quality of the urban environment, it
set arbitrary design, and the cities became "procrustean bed"[22] of trees.

In the 18th century, a new style of embellishment and aesthetic has
emerged; British expanded cities over the walls of the Middle Age, including
trees within open areas of new districts and an imitation of the natural
landscape, became common in the design of parks, squares, and residential
properties (Lawrence, 2008). *Homo aestheticus* experimented new vision of
aesthetic based on "natural beauty *per se*". The picturesque[23] and *Genius
loci*[24] concepts drew attention to the hidden beauty in the spirit of the place
and wilderness (Selman and Paul, 2010).

Under embellishment and aesthetic, the "wilderness" returned to the
world of *civis*. Not only trees were unpruned for aesthetic purpose, but also
"wilderness" became beautiful. Common historical roots for the modern idea
of *Wild Urban Woodlands*[25] can be drawn to the aesthetic idea of wilderness
in the 18th century (Lawrence, 1993).

---

[22] In the book The *Purloined Letter*, Edgar Allan Poe (1844) used the metaphor of
"Procrustean bed" to describe rigid method of Parisian police. (Procrustean bed refers to
Prokoptas or Damastes in Greek myth, who put travellers in his bed, stretched, or lopped
off their limbs to adapt them to its length. Tripp, Edward, T. (1970). *The Meridian Handbook
of Classical Mythology.* Meridian, p. 498.)

[23] See also Wordsworth, W. (1810). *Guide to the Lakes*; it reflects on picturesque sentiments
of natural features.

[24] See also Bell, S. (2012). *Landscape: Pattern, Perception and Process.* 2nd Edition.
Abington, New York: Routledge; it mentions to the contributions of Schopenhauer and natural
beauty relates to *Genius loci*. In Italian version, Norberg-Schulz, C., Norberg-Schulz A.M.
(1992). *Genius Loci. Paesaggio, ambiente, architettura, Documenti di architettura.* Milano:
Electa.

[25] Tile of the book: Kowarik, I. and Körner, S. (eds.) (2005). *Wild Urban Woodlands. New
Perspectives for Urban Forestry.* Berlin: Springer.

3.4 Power and Luxury

People with social and political power expressed their supremacy by planting trees in or around their houses. In fact, from Chinese gardens to European gardens, a small number of prosperous people enjoyed the aesthetic of nature in their private residences. Access to these private gardens was permitted to the minority of wealthy people with high social status. For instance, in Europe, in the 16th century, some private garden was open to the public through the gate on which was posted *Lex Hortorum* (the law of gardens) to specify necessary manner to frequent the garden (Coffin, 1982); Chinese garden was opened to well-dressed foreigners and Chinese were refused to enter (Zimmerli, 2010).

This segregation was gradually eased, and some gardens of monarchs and kings were opened to the public and a more comprehensive range of social classes. We can mention Tuileries, Tiergarten, and Hyde Parks that later became the first large parks in the urban context. Other new public spaces like the *London Pleasure garden* and *Jardis spectacles*[26] of Paris have appeared for people with money and time to enjoy *locus amoenus*. Such theatrical entertainment of public space has been described by the *flâneur*'s[27] vision of the city. Such discrimination finally abandoned when the sense of *noblesse oblige* rised among wealthy commoners who used their private fund to create public space and plant trees for the public's benefit.

Nonetheless, the presence of trees near residence expressed the sense of luxury, and, even today, the posh vicinities of cities benefit from more green space than poverty-stricken neighbourhoods. The historical roots of this phenomenon can be drawn to the estate developers and landlords, who included trees in their development schemes as "profit-oriented civic philanthropy" (Lawrence, 2008). The creation of large parks of the 19th century followed the same logic; increasing the value of properties and generating tax profit was the base of these parks, which led to the imbalanced distribution of trees in urban contexts. For instance, the terrain of Central Park of New York was chosen to evict the Afro-America squatters of a rocky, swampy area. Since this terrain was not suitable for construction, it has been

---

[26]See also Panzini (1993). *Per I piaceri del popolo. L'evoluzione del giardino pubblico in Europa dale origini al XX secolo;* Langlois G.A.(1991). *Folies Tivolis et attractions. Les premiers parcs de loisirs parisiens,* Délégation à l'action artistique de la ville de Paris; Isherwood, R.M. (1986). *Farce and fantasy. Popular entertainment in eighteenth-century Paris.*

[27]*Flâneur* is the urban adventurer who enjoyed walking the streets for the pleasure received by the city life, invented by Charles Baudelaire (1857) *Les Fleurs du mal.*

adopted by planners as park for wealthy people of uptown (Rosenzweig and Blackmar, 1992).

To conclude, the history of trees in cities is an anthropocentric expression of power, which has led to social segregation, widening the gap between poor and rich.

From the present perspective, the power of nature[28] has generated "global" discrimination and segregated poor societies from rich countries. Regarding such discrimination, Vandana Shiva (2005) writes:

> "If we are serious about ending poverty, we have to be serious about ending the systems for wealth creation which create poverty by robbing the poor of their resources, livelihoods and incomes. Before we can make poverty history, we need to get the history of poverty right. It is not about how much more we can give, so much as how much less we can take".

The system of wealth accumulation is based on riches taken from the poor, which caused poverty and inequity and resulted in the environmental crisis of our era. For instance, deforestation due to the export of beef, palm oil, soy, and wood products, including timber and paper, has resulted in environmental disaster in countries like Brazil, Indonesia, Malaysia, the Democratic Republic of the Congo, Papua New Guinea, Bolivia, Argentina, and Paraguay (Persson et al., 2014).

3.5 Towards the Age of Enlightenment

> "The more you think about the services of the Forest, the more you understand them, the more essential they appear. It is true indeed that the Forest, rightly handled – given the chance – is, next to the earth itself, the most useful servant of man".[29]

> Gifford Pinchot

With the burgeoning population and city expansion, a new idea of optimal city has emerged, and the simple idea of embellishment and aesthetic shifted

---

[28] See Joachim Radkau (2008) *Nature and Power: A Global History of the Environment.* New York: Cambridge University Press.

[29] Pinchot. (1998). *Breaking New Ground.* Washington, D.C.: Island Press, p. 32. Pinchot was the founder and first chief of the Forest Service 1905-1910 in USA.

to enlightenment. Furthermore, the modern conservation movement[30] opened a vision towards ecological awareness and the protection of natural resources.

"Enlightened Homo" perceived trees not only as a resource of beauty (English garden) but also as prosperity to improve the cleanliness of urban space, purify polluted air, and setting public space for individual health (Lawrence, 2008). The idea of the garden city movement[31] and open urban design initiated with enlightenment. Patrick Geddes was the first and most influential in constructing the city in a regional vision with advocacy towards nature. He discussed "how the residents' use of a city influenced its form"; moreover, he introduced the "Geddes garden", where the production of food and plants was aimed for educational purpose (Günter Voss, 2008).

Cities expanded to the far suburban landscape, and new suburban houses arose in several places as the anti-urban aesthetic. The first emergent difference seems to be the large size of trees and the use of exotic plants. In the United States, people personalized the space in front of their house by planting exotic trees as memories of their homelands. The new use of trees in the United States developed a new form of green space, and large city parks have planned. Fredrick Law Olmsted, H.W.S. Cleveland,[32] and Charles Eliot[33] designed parks to incorporate the trees into the city.

As an interdisciplinary approach, the history of actual urban forestry was initiated with the collaboration of Olmsted and Gifford Pinchot,[34] who linked traditional forestry to urban park disciplines. Roxi Thoren (2014) explains that "Where most Americans saw a conflict between forest conservation and timber extraction, Olmsted and Pinchot saw the two as coexisting — healthy forests could produce useful timber [...] Biltmore forests [as an

---

[30]The first conservation movement returned back to 1662, when John Evelyn's treatise, *Sylva,* proposed the idea of conservation of forests by controlling the rate of depletion and plantation. (Nisbet, 2007) [online] Available at: http://www.gutenberg.org/files/20778/20778-h/20778-h.htm) [Accessed 13 August 2017].

[31]The Garden City concept proposed by Ebenezer Howard (1898); he intertwined social and economic concerns with physical environment and advocated more open urban design.

[32]Cleveland worked mainly in shaping the park system and waterfront of Minneapolis and St. Paul.

[33]Charles Eliot had a main role for the park system of Boston. He is the writer of *Garden and Forest* (1897), illustrating the need to manage forests and not simply protect and preserve them.

[34]He was (at the time) the only US-born, trained forester who was educated in Europe (France, Germany, and Swiss) in the scientific practice of forest management; although young, he was one of the nation's leading experts on forestry (Thoren, 2014).

arboriculture laboratory followed three goals:] to generate profit; be self-sustaining; and improve the health of the forest [...] materially, economically, and ecologically productive".

The collaboration of Olmsted and Pinchot is an example of "Enlightened" Anthropocentrism[35] based on complete comprehension of the problems involved. They interlaced the utilitarian philosophy with the conservation and protection of natural resources. They illustrate a novel vision of the use of forest in which the previous pure anthropocentric idea of utility shifted towards environmental conservation in which there is a "state of harmony between men and land" (Leopold and Schwartz, 1949).

Enlightened Anthropocentrism brought forests into cities and "adopted principles of forestry [...] to achieve and maintain a balanced age structure within the urban locality to ensure continuous tree cover, and hence sustained provision of goods and services" (Konijnendijk, 2003). The efforts of Olmsted for scientific land management have essentially contributed to the new paradigm, ecocentrism—a turning point for overcoming the duality between city and nature. He was "a great pre-professional, synthetic designer who knew no professional boundaries and whose projects hybridized aesthetics, infrastructure, ecology, social justice,[36] and economic productivity" (Thoren, 2014). His legacy goes far beyond his masterworks and has bequeathed a set of principles that have served as a blueprint for sustainable design and environmental conservation.[37]

---

[35]Enlightened Anthropocentrism in philosophy explains that "humans do have ethical obligations toward the environment, but they can be justified in terms of obligations toward other humans", and "it is viewed as immoral, because it deprives future generations of those resources". Sustainable development has common roots in Enlightened Anthropocentrism. See Anthropocentrism, by Sarah E. Boslaugh [online] *Encyclopedia Britannica.* Available at:https://www.britannica.com/topic/anthropocentrism#ref1187034 [Accessed 14 August 2017].

[36]It is readable from his writing that he was democratizing nature. Best voiced by Olmsted in his writings about Central Park is written that "It is one great purpose of the Park to supply to the hundreds of thousands of tired workers, who have no opportunity to spend their summers in the country, a specimen of God's handiwork that shall be to them, inexpensively, what a month of two in the White Mountains or the Adirondacks is, at great cost, to those in easier circumstances".

[37]See National Association for Olmsted Parks, Available at: http://www.olmsted.org/the-olmsted-legacy/olmsted-theory-and-design-principles/design-principles [Accessed 30 December 2017].

Colonized Forests
Conclusion

In conclusion, this chapter reviewed the duality between city and forest, with a historical background. Such duality arose because of anthropocentric vision, in which civilized human, as "Ego" above nature, planted trees in the cities to express their "supremacy" and to satisfy their sense of "ego".

First, the forests were separated from cities and depleted due to the sacredness. Then, the forests were exploited and extracted as resources. Later, trees were imported to the city, isolated in the gardens, sacrificed for utility, decoration, and expression of power, provoking pleasure and aesthetic. Finally, enlightenment introduced a prosperous idea of the city that was based on the benefits of trees.

Nonetheless, the cities are found on Anthropocentrism, and the forests are caged in cities. Urban forests such as Golden Gate Park and Central Park have emerged with the city's development, and the solid figural forms of them reveal the imposed organizational system of the city (Czerniak, 2007). Such antithetical archetypes have been enforced to coexist depending on the context, and as Misharina and Chung (2014) accentuate:

> "The Forest caged in the city is swallowed by it, digested. It can never remain the same. It has to live by the rules set in the city; it has to survive and adapt. It has to wear masks that are for the city embedded faces. It has to hide; it has to suffer. It is colonized".

If we perceive the cities in another vision, which is not anthropocentric, if we see human not above nature but as a part of it, and if we study the influence of our cities on a broader scale, how will urban forests be? The next chapter discusses these questions basing on holistic theories.

References

Barrow, E.G.C. (2010). Falling between the 'Cracks' of Conservation and Religion: The Role of Stewardship for Sacred Trees and Groves. In: Verschuuren, B., Wild, R., McNeely, J. Oviedo, G.(eds.) *Sacred Natural Sites Conserving Nature and Culture.* London, Washington D.C., Gland: Earthscan, pp. 42-52.

Bacon, F. (1620). *Novum Organum.* Available at: http://www.constitution.org/bacon/nov_org.htm [Accessed 14 August 2017]

Brosse, J. (1989). *Mythologie des arbres.* Paris: Payot.

Catton, W. and Dunlap, R. (1978). Environmental Sociology: A new paradigm. *The American Sociologist*, 13. pp. 41-49.

Chadwick, L.C. (1971). 3000 Years of Arboriculture: Pat, Present and Future. *Arborist News*, 36(6), p. 73-78.

Chung, K., Misharina, A. (2014). Paradoxes of Archetypes: The Urban and the Forest. [online] Scenario 04: Building the Urban Forest. Available at: https://scenariojournal.com/article/paradoxes-of-archetypes/ [Accessed 14 August 2017]

Coffin, D.R. (1982). The 'Lex Hortorum' and access to gardens of Latium during the Renaissance. Journal of garden history 2(3), p. 201-232.

Czerniak, J., Hargreaves, G.(eds.) (2007). *Large parks*. New York: Princeton Architectural Press. p. 26.

Evernden, N. (1984). The environmentalist's Dilemma. In: Evernden, N.(ed.) *The Paradox of Environmentalism*. Faculty of Environmental Studies, Toronto: York University, p. 7-17.

Foreman, D. (1991). *Confessions of an eco-warrior.* New York: Crown publishers.

Freud, S. (1927). *The Future of an Illusion.* cited in Rodes, B and Rice Odell, R. (1992). *A Dictionary of Environmental Quotations*. New York: Simon & Schuster, p. 197.

Gerhold, H.D., Frank, S.A. (2002). *Our Heritage of Community Trees*. Mechanicsburg: Pennsylvania Urban and Community Forestry Council.

Grimal, P. (1974). *L'arte des Jardins*. Trans. Magi, M. Roma: Feltrinelli

Günter Voss, R. (2010). Temporary Tree Nurseries. In: Ghiggi, D. (ed.). *Tree Nurseries Cultivating the Urban Jungle: Plant production Worldwide*. Baden: Lard Müller publisher.

Hareuveni, N. (1980). *Nature in our Biblical Heritage*. Kiryatone: Neot Kedumin.

Harmon, D. and Putney, A.D. (2003). Intangible values and protected areas: Towards a more holistic approach to management. In: Harmon, D. and Putney A.D. (eds.). *The Full Value of Parks: From Economics to the Intangible*. New York: Rowman and Littlefield, p. 311–326.

Harrison, R.P. (1992). *Forests: the shadow of civilization*. Chicago: University of Chicago Press.

Khan, K.H. (1999). An Islamic perspective on the environment. In: Dempsey, C.J. and Butkus, R.A. (eds.) *All Creation is Groaning: An Interdisciplinary Vision for Life in a Sacred Universe*. Minnesota: Liturgical Press, p. 46–57.

Konijnendijk C.C. (2003). A decade of urban forestry in Europe. *Forest Policy and Economics* 5, p. 173-186.

Lawrence H.W. (1993). The Neoclassical origins of modern urban forests. *Forest and Conservation History* 37(1), p. 26–36.

Lawrence, H.W. (2008). *City Trees: A Historical Geography from the Renaissance through the Nineteenth Century*. Charlottesville: University of Virginia Press.

Leopold, A. & Schwartz, C. W. (1949). *A Sand County Almanac, and Sketches here and there*. New York: Oxford Univ.

Livingston, J.A. (1985). Moral concern and the Ecosphere. *Alternatives*, 12(2), p. 3-9.

Miller, R.W., Hauer, R.J., Werner, L.P. (2015). *Urban Forestry: Planning and Managing Urban Greenspaces*. (Third Edition). Long Grove: Waveland Press, p. 44-45

Nasr, S.H. (1998). *The Spiritual and Religious Dimensions of the Environmental Crisis*. London: Temenos Academy.

Oelschlaeger, M. (1991). *The Idea of Wilderness: From Prehistory to the Age of Ecology*. New York: Yale University Press. p. 61, p. 286.

Oviedo, G. and Jeanrenaud, S. (2007). Protecting sacred natural sites of indigenous and traditional peoples. In: Mallarach, J.M. and Papayannis, T. (eds.) *Protected Areas and Spirituality*. Gland: IUCN and Publicacions de l'Abadia de Montserrat.

Panzini, F. (1993). *Per i piaceri del popolo. L'evoluzione del giardino pubblico in Europa dalle origini al XX secolo*. Bologna: Zanichelli Editore.

Persson, M., Henders, S. Kastner, T. (2014). Trading Forests: Quantifying the Contribution of Global Commodity Markets to Emissions from Tropical Deforestation. *CGD Working Paper* 384. Washington, DC: Center for Global Development. [Online] Available at: https://www.cgdev.org/pu blication/trading-forests-quantifying-contribution-global-commodity-ma rkets-emissions-tropical [Accessed 12 August 2017]

Profuse, G.V. (1992). Trees and urban forestry in Beijing, China. *Journal of Arboriculture,* 18(3), p. 145-153.

Ramakrishnan, P.S. (1996). Conserving the sacred: From species to landscapes. *Nature and Resources*, 32(1), p. 11–19.

Rosenzweig, R., Blackmar, E. (1992). *The Park and the People: A History of Central Park*. New York: Cornell University Press.

Schneckenburger, S. (2001). *Carl Albrecht Purpus (1851-1941) - Ein deutscher Pflanzensammler in Amerika*. Freundeskreis des Botanischen Gartens der TU Darmstadt.

Selman, P. and Swanwick, C. (2010). On the Meaning of Natural Beauty in Landscape Legislation. [online] *Landscape Research*, 35(1). Available at: [Accessed 11August 2017]

Shiva, V. (2005). Two myths that keep the World poor. *Ode 28*. Available at: http://eprints.whiterose.ac.uk/11121/2/selmanp_natural_beauty_paper.pdf [Accessed 19August 2017]

Sitte, C. *City planning According to Artistic Principals*. Trans. Collins, G.R., Collins, C.C. (1965). New York: Random house, p. 176, 178, 167-185.

Thoren, R. (2014). Deep Roots: Foundations of Forestry in American Landscape Architecture. [online] Scenario 04: Building the Urban Forest. Available at: https://scenariojournal.com/article/deep-roots/ [Accessed 13 August 2017]

Voltaire. (1749). "Des Embellissemens de Paris". In: Waddivor, M. (ed.) (1994) *The complete Works of Voltair*, 31B, p.199-233. Oxford: The Voltaire Foundation.

Waisel, Y. and Alon, A. (1980). *Trees of the Land of Israel*. Tel Aviv: Division of Ecology.

Williams, M. (2006). *Deforesting the Earth: From Prehistory to Global Crisis*. Abridged edition. Chicago: University of Chicago Press, p. 149.

Zimmeli, C. (2010). Better city, Better Life. In: Ghiggi, D. (ed.). *Tree Nurseries Cultivating the Urban Jungle: Plant production Worldwide*. Baden: Lard Müller publisher.

4

Holism: Forests in Cities/Cities in Forests

Forests in cities/cities in forests

> "A thing is right when it tends to preserve the integrity, stability, and beauty of the biotic community. It is wrong when it tends otherwise".

Aldo Starker Leopold[1]

In the face of environmental awareness, the integrity of ecosystems and the survival of all species became central debates of the new era. The previous dualistic and reductionist vision of the world shifted to a holistic view. Interdisciplinary approaches have emerged to resolve the complexity of environmental issues.

As the necrotic result of Anthropocentrism and in the crisis of the urban age, the city draws the attention of all scientists, ecologists, philosophers, planners, geologist, architects, and landscape planners. The school of thoughts of Olmsted has played a crucial role, widening debates towards novel humanity as the driver agent to hybridize aesthetics, ecology, and well-being to the whole system. This chapter goes through a different holistic approach, from local to a regional and global scale, to figure out in what extension the urban forests can operate, in line with the emergence of "global city".

The section evaluates the hypothesis of "global forest" based on the urbanization process in a holistic vision of global issues.

---

[1] Aldo Leopold introduced the concept of ecocentrism in the book *A Sand County Almanac* (1949); his report of Yellowstone national park is officially known for ecosystem management.

4.1 From Anthropocentrism to Ecocentrism

From the middle of the 20th century onwards, in the face of the ecological apocalypse, humanity's vision of the world shifted *"from Ego to Eco"*.[2] Ego refers to the well-being of self—Anthropocentrism—while eco (or *"oikos"* in Greek means "the whole house") indicates holism—the well-being of the whole system. Scharmer and Käeufer(2013) make a clear description about this paradigm shift:

> "It is shift from an *ego*-system awareness that cares about the well-being of oneself to an *eco*system awareness that cares about the well-being of all, including oneself. When operating with ego-system awareness, we are driven by the concerns and intentions of our small ego self. When operating with ecosystem awareness, we are driven by the concerns and intentions of our emerging or essential self— that is, by a concern that is informed by the well-being of the whole".

According to this paradigm shift, modern environmental ethics[3] and conservation movements have risen. Eugene Odum, one of the most pioneering idealists of holism, drew attention to the integrity of ecosystem ecology, emphasizing the well-being and stability of the planetary ecosystem as "the basic unit of structure and function".[4] Holistic vision has led to interdisciplinary perspectives, and George Bateson (1979), as one of the promotors of such perspective, called attention to the interaction of systems:

> "We have been trained to think of patterns, with the exception of those of music, as fixed affairs. It is easier and lazier that way but, of course, all nonsense. In truth, the right way to begin to think about the pattern which connects is to think of it

---

[2] *"from Ego to Eco economy"* is a research program conducted by Presencing Institute (PI) and supervised by Scharmer and Käeufer. It has been published as a book in 2013, in which they discuss socioeconomic challenges, proposing methods for transforming capitalism.

[3] Environmental ethics based on nature-centered; under this paradigm, new theories like ecocentrism, biocentrism, environmental design, human ecology, and sustainability have emerged. For the history of environmental ethics, see Cramer, P.F. (1998). *Deep Environmental Politics: The Role of Radical Environmentalism in Crafting American Environmental Policy.* Westport: Praeger; Nash, R. (1990). *The Rights of Nature: History of Environmental Ethics (History of American Thought & Culture).* Madison: University of Wisconsin Press.

[4] See also Noer Lie, S.A. (2016). *Philosophy of Nature: Rethinking Naturalness.* p. 193.

as primarily (whatever that means) a dance of interacting parts and only secondarily pegged down by various sorts of physical limits and by those limits which organisms characteristically impose".

In the era of ecocentrism, as the opposite of Anthropocentrism, novel forms of utopia have been planned to reduce the impact of "habitat" of *civic* on the planet and designated the role of "stewardship"[5] to *civic* (McHarg, 1992). The forests became a competent tool for creating such green utopias that "is sustainable only if the well-being of the world's ecology is a part of the plan" (Marshall, 2016). In terms of ecocentrism, the cities are not created just for human well-being, but they contribute to the well-being of the whole system; cities, forest, human, and plants become whole and if "we try to pick out anything by itself we find it hitched to everything else in the Universe"[6] (Muir, 1911). Cities are not systems *per se* but "offer arenas for enormous opportunities [7] [...] with room for humans and other species to thrive" (Elmqvist et al., 2013).

The idea of a *Biophilic*[8] city is a prevalent prototype, which is based on ecocentrism. A Biophilic city recognizes "the innately emotional affiliating of human beings to other living organisms" (Wilson, 1984) and "puts nature first in its design, planning and management" (Beatley, 2016). If a Biophilic city is an excellent example of ecotopia, what shape do the "utopian forests" have to include human and biodiversity? How do they influence on the whole system? To find an answer to the mentioned questions, the chapter follows with different holistic approaches of forestation and urbanization, which have roots in landscape ecology development as one of the approaches of conservation movement (Forman, 2015).

---

[5]McHarg emphasized on the role of human as Steward for conservation of ecosystem. See Richard Weller (2014) Stewardship Now? Reflections on Landscape Architecture's Raison d'être in the 21st Century. *Landscape Journal, 33(2),* p. 1-24.

[6]Cited by Giffor, T. (2006). *Reconnecting with John Muir: Essays in Post-Pastoral Practice. Georgia:* University of Georgia, p. 25.

[7]Global Assessment of Urbanization, Biodiversity and Ecosystems Services report accentuates that due to the urbanization and expansion of cities, cities include wide range of biodiversity; thus, they are also increasingly recognized for their role in conserving biodiversity.

[8]The concept of *Biophilia* originally was coined by German social psychologist Erich Fromm. Cited by Beatly, T. (2015).

4.2 Optimal Form of Forests

During the 20th century, holism gives birth to novel precursors of greening; Green Belt of London, Green infrastructure, Ecological Network in northern EU countries, Green System in Balkan area, Green Way in North America, Biodiversity Corridors in the Philippines, and Urban Greening in Singapore (Fábos and Ryan, 2004). These are heterogeneous terminology to define the integrity and connection of these green structures that create a whole, and the urban forests are a part. Greenbelt, Greenway, Green Infrastructure, and Ecological Network evolution is the main concern of this section to point out the primary directive and challenge that these holistic approaches have encountered.

In 1935, Green Belt of London prevented the urban sprawl of the city. The openness of land obtained this purpose in surrounding areas of the city. The main disadvantages attributed to Green Belt are: lack of protection of high biodiversity land (for instance, suburban gardens are not incorporated to Green Belt) (Davidson and Wibberley, 1977); Green Belt policy led to higher densities and fewer gardens in suburban areas (Williams, 2000), and social justice is another shortfall; restricting supply and raising land values within constrained urban areas led to a shortage of affordable housing (Amati and Taylor, 2010). Without forests, the Green Belt is not *de facto* green and productive (Konijnendijk, 2010).

Later in 1987, Greenway in North America has been proposed to create networks of linear elements for multiple purposes based on sustainable land use policy.[9] The opponents of Greenway remark the ecological aspects such as inefficient use of corridors for many species which disperse across the landscape without corridors or the spread of invasive species into protected areas (Daniels, 1988). Furthermore, Ahern (1995) mentions that "imposing greenway corridors in a landscape may lead to greater uniformity and a loss of cultural landscape identity. In open landscapes, forested corridors are an unnatural and inappropriate introduction, which can radically change the landscape physically, culturally and visually".

In line with the development of concepts in North America, in Europe, French landscape designers like Jaques Simon and Michel Corajoud blurred boundaries between disciplines; ecological design has been adopted as the part of plans in urban and suburban areas and other concerns like

---

[9]Based on literature review of greenways in landscape planning (President's Commission on Americans Outdoors, 1987; Little, 1990; Flink and Seams, 1993; Smith and Hellmund, 1993).

préverdissement (or forestation before the urbanization) has been introduced by Guinaudeau. Others like Michel Desvigne and Christine Dalnoky proposed the rediscovery of a broader scale of territory to face ecological challenges, conceiving a new type of relationship between urban and the surrounding countryside. Several projects like Thomason Factory at Guyancourt[10] and the TGV station at Avignon reveal novel prototypes of forestation related to broader scales and issues. Gilles Clément in Parc Henri Matisse in Lille advocates the metaphoric forest of L'île Derborence as a myth of nature and as a manifest to overcome the dualism between the natural world and urban areas (Metta, 2008).

In 1994, Green Infrastructure has been proposed as a land conservation strategy with the notion that natural systems are equally, if not more important, components of our "infrastructure" (Firehock, 2010), providing an adequate ecosystem that maintains natural ecological processes, sustains air and water resources, and contributes to the health and quality of life (Benedict and McMahon, 2006). The central debates against Green Infrastructure (GI) refer to the grey infrastructures, which are highly clogged with traffic, and beneficial effects of GI would be faded away (Socco et al., 2007).

To summarize, the main aim of these holistic strategies is to tackle the fragmentation in the landscape due to urbanization. Spatial connectivity is fundamental for the long-term survival of many species,[11] albeit some species are better left in isolation.[12]

Creating connectivity out of the urbanized area is more achievable; but in urban areas, is connectivity the "correct" and "possible" answer? What is the optimum shape? Can we embody an ideological form? Thinking holistically implies a considerable amount of vision and commitment. The diversity and complexity of ecology and traditional urban form pose challenges for defining sustainable design for urban forests. Lister (2007) stresses that we cannot define a "correct" state in ecological design. The habitat of some species could create conflicting habitat for the other species; both habitats are valid but not at once.[13] So, we can define an "adaptive ecological design" with the

---

[10]Commissioned by Renzo Piano.

[11]European Program of LIFE is based on GI to connect the habitats of endangered species to assist genetic exchange, including stopover areas (for instance, migrating birds) and corridors linking habitats. Since 1992, LIFE has co-financed some 4306 projects of conservations.

[12]See also debates regarding SLOSS: protecting single-large-or-several-small patches or ecosystems.

[13]She refers to this example to explain the correctness in ecology: "Fast flowing streams that support trout spawning may eventually become stagnant warm-water ponds if beavers are

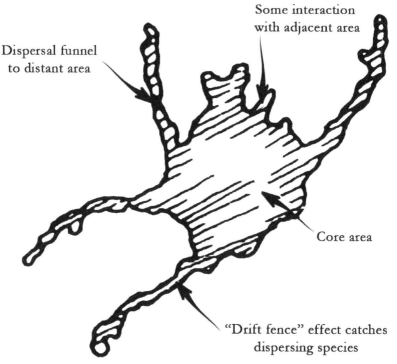

Figure 4.1 Author's re-elaboration of ideal patch shape by R. Froman.[14]

capacity for resilience and recovery from disturbance. However, in terms of ecological function, Richard Forman (1986) suggests "spaceship-shaped" as an optimal form for a patch (Figure 4.1), which can be imagined as an urban forest, with a rounded core for protection of resources, plus some curvilinear boundaries and a few fingers for species dispersal.

The "connectivity" of mentioned green infrastructure, the "optimum ecological form" of the patch, and "adaptive ecological design" should be the main concerns for designing a sustainable urban forest, which can contribute to the stability of the whole entity over time as Odum suggested.[15]

---

allowed to do their work. The trout will die out while the beaver flourish". There are similar debates according to Odum theory of stability of system. See the influence of wolves in Yellowstone National Park on forests: Peter Wohlleben. (2015). *The Hidden Life of Trees.*

[14] See Dramstad E.W., Olson, J.D., Forman, R.R. (1986). *Landscape Ecology Principals in Landscape Architecture and Land-use Planning.* p. 32.

[15] Refers to "nature is as a whole in a steady state or is in the most stable form possible and constitutes one big entity [...] a system which has stability with time will exist

In a holistic approach, which is the base of ecotopia, the design of urban forest does not emerge from common imposed forms of the city but comes up as the result of ecological function and evolution, implying an "open-ended process [where] final evolving form is not defined" (Hargreaves, 2007) and the forest is *en movement* (Clément, 1991). Such ecological design helps to overcome the dualism between urban and forests (nature and culture), and as Clément stresses, it aims "to redefine the theory about the place of man in nature".

In such a view, the novel theory of holism, in which the human power over nature is reversed and the idea of resourcism is comparable to "ecosystem services",[16] the issues of embellishment is assimilated to "wilderness"[17] and its *beau désordre... cause notre admiration.*[18] The previous regular form of forest shifts to an endless irregular forest that passes through the cities linking with other entity; such a new ecological vision of forestation brings to mind Italian writer Calvino's evocative image of the forest in which "the monkey who had left Rome, skipped from tree to tree, till it reached Spain, without ever touching the ground".[19]

The optimal ecological form of forests is clear; however, how is it applicable to the cities? What is the city? What is "the boundary" of urban forests? Is it just a matter of connectivity?

---

longer than a system without stability". Odum, H.T. (1950). Some Biological Aspects of the Strontium Cycle Introducing a new Tool of Paleoecology. In U.S. Science holds its biggest powwow - Reports on some of the year's big discoveries. *Life Magazine,* Jan. 9, 1950, p. 8-9.

[16]It has been recognized the "natural capital" of forests by economists, in different reports like Millennium Assessment Reports, The Economics of Ecosystems and Biodiversity (TEEB), and Earth Economics.

[17]See Kowarik, I. and Körner, S.(eds.) (2005). *Wild Urban Woodlands. New Perspectives for Urban Forestry.* Berlin: Springer.

[18]From original test: *Cette expérience (dans le jardin de la Vallée) a duré huit ans, au terme desquels j'ai réalisé qu'il s'agissait d'une nouvelle forme de jardinage mais aussi d'une théorie visant à redéfinir la place de l'homme dans la nature.* Gilles Clément, "Libérez les jardins", Le *Nouvel Observateur,* 1998, cited by Danielle Langeais, "The garden in movement: ecological rhetoric in support of gardening practice", *Studies in the History of Gardens & Designed Landscapes,* vol. XXIV, no. 4, 2004, p. 313-340, ici p. 336, note 29.

[19]Calvino, I. (1977). *The Baron in the Trees.* Mariner Books. See also Di Carlo, F. (2013). *Paesaggi di Calvino.* Melfi: Libria.

4.3 Forest and Cities in Avant-garde Planning Approach

> "The ecological crisis is doing what no other crisis in history has ever done - challenging us to a realization of a new humanity".

Jean Houston

In the Anthropocene era, humanity and nature became one and embedded[20]; thus, the survival of "new humanity" is rooted in the durability of "new nature". The era of Anthropocene called a paradigm shift in science, asking new model of management. If it is just "three years left to save the planet",[21] how humanity can stop "the end of the world?"[22]

Amidst the crisis, landscape as "the result of action and interaction of natural and human factors"[23] assumes a vital duty. Holistic theory of landscape found on hierarchal systems view of the world, rooted in interdisciplinary approaches, can solve the complexity and duality between nature and culture (forest and city) (Naveh, 2001). Maybe we cannot act on the sky and the light; but on the rest of the landscape, we have all the direct responsibilities derived from our acts and omissions[24] (Di Carlo, 2016). The management of city as human ecosystems (Steiner, 2011) and their relation to other systems have become the primary responsibility of landscape architecture during the last decades. This section goes through the history of planning approaches to determine whether prior proposals were/are efficient and productive for the 21st century.

In the holistic vision of the city, experts attempted to intertwine urbanism with landscape and ecology to overcome the gap between city and nature. The

---

[20]From the test of: "Nature as we know it is a concept that belongs to the past. No longer a force separates from and ambivalent to human activity, nature is neither an obstacle nor a harmonious other. Humanity forms nature. Humanity finds itself embedded within the recent geological record". Haus der Kulturen der Welt. (2013). [online]. *The Anthropocene Project: An opening (10-13 January 2013)*. Available at: www.hkw.de.

[21]Referring to the letter of scientist published in the journal *Nature* (July 2017) about the importance of next three years to stop the global warning.

[22]See Atlas for the End of the World, conservation of endangered bioregion on Earth and status of land use and urbanization, by Richard J. Weller, Claire Hoch, and Chieh Huang. *Atlas for the End of the World* (2017), Available at: http://atlas-for-the-end-of-the-world.com.

[23]Definition of landscape, European Landscape Convention, Florence, 2000.

[24]Original Italian passage: "Se sul cielo e la sua luce possiamo agire poco, sul resto del paesaggio abbiamo tutti delle responsabilità dirette, che derivano dai nostri atti e omissioni" (Di Carlo, 2016, p. 15).

ecological advocacy of McHarg became the base of these planning methods. Charles Waldheim (2006) coined the term "landscape urbanism" integrating landscape to urban design, stressing the role of "Landscape Architects as urbanists of our age" (Waldheim, 2017); his concern deals with urban scale-landscape as the "building block for urban design" (Steiner, 2011).

Forman and Gordon (1986) identified the field of landscape ecology to illustrate its promising potentiality for regional planning. Steiner (2014) makes a synthesis of all proposals, stating Landscape Ecological Urbanism with the "goal to design and plan cities to increase, rather than to decrease, ecosystem services".

The necessity of the interdisciplinarity approach with interconnection scales, from urban to regional, is accentuated by all experts. Similarly, Franco Zagari highlights (2013) that there is always mutual contamination and a logical continuity; for instance, the garden as a laboratory of ways of being and living behaviours prompts new forms of organization for the territory and the city.[25]

Nonetheless, such a statement is mainly theoretical and highly discarded from projects. For instance, planning pursues an ecological network and mainly focuses on the areas outside of cities and vice versa, designing emphases on parks, gardens, and public urban spaces in cities. This segregation between large and small scales, planning and designing, is readable in most greening projects.

Foresters make a similar separation between forests according to geographical limits, putting adjectives like "urban", "suburban", and "rural". Cecil Konijnendijk et al. (2006) attempted to define an international terminology for urban forestry.[26] As a result, it has been highlighted that since the "traditional dichotomy between city and countryside is no longer very real, defining geographical limits" are hard to draw. Nonetheless, it has been expressed support for "a broad and holistic definition that incorporates ecological, economic, and sociological elements of 'urban forest' and is inclusive of people from cities to suburbs to rural communities". Today, this notion is mainly debated as ecosystem services, to balance the broader definition with managerial approaches.

---

[25]Original Italian passage: "Fra le varie dimensioni vi è sempre una contaminazione reciproca e una continuità logica, che vede ad esempio il giardino come laboratorio di modi di essere e comportamenti dell'abitare che sperimentano nuove forme di organizzazione del territorio e della città" (Zagari, 2013 p. 47).

[26]Note that *urban forestry* is management of urban forests.

©Fabio Di Carlo—Costiera Amalfitana.

The historian, Henry Lawrence (2008), at the end of the book, *City Trees*, highlighted that:

> "[...] the most important difference today is the enormous physical scale of cities. The term "city" has lost much of its meaning. They are metropolitan areas, even metropolitan regions

[...] The zone of their influence spreads perceptibly [...] How does one speak on New York City when its metropolitan influence extends from central Connecticut to the Poconos in Pennsylvania? Is the mayor of New York City more responsible for what happens to trees in this area than the governor of New Jersey?"

Who is responsible, and to what extent? Who must plan and design the "urban" forests, and to what extent? Which method is required? Designing or planning? Local scale, regional scale, or global scale?

4.4 Novel Boundaries for Forests and Cities in a Global Vision

"The city is everywhere and everything. If the urbanized world now is a chain of metropolitan areas connected by places/corridors of communication (airports and airways, stations and railways, parking lots and motorways, teleports and information highways), then what is not urban?

Is it the town, the village, the countryside? Maybe, but not to a limited degree. The footprints of the city are all over these places, in the form of city commuters, tourists, teleworking, the media, and the urbanization of lifestyles. The traditional divide between the city and the countryside has been perforated".

Amin and Thrift (2002)

In the era of the Anthropocene, humanity should also encounter the urban age.[27] More than any previous time, the steady growth of population, over half lives in cities,[28] hovers over humankind new challenges. Outside of some portions of inhospitable land, much of the globe has been urbanized; "visible 'urban tsunami' sweeps swiftly and powerfully across the land" (Forman, 2008). The advanced hypothesis of Henri Lefebvre in 1970, "the complete urbanization of society", expressed that the process of urbanization

---

[27]London School of Economics program "Urban Age" http://lsecities.net/ua/. This research analyses the performance of cities based on a comparative data on spatial, social, economic, and environmental drivers.

[28]See Population Reference Bureau; By 2030, 60% of population will live in urban area. Available at: http://www.prb.org/Multimedia/Video/2012/distilled-demographics-urbanization.aspx

creates the conditions for capitalism, which is the reason for the *implosion and explosion* [29] of capitalist urbanization. It has also emerged the question of *Anthropocene or Capitalocene?* (Moore, 2016)

The expansion of cities generates an ecological footprint and leads to land consumption, deteriorating the environment. Although all these crises necessitate halting further urbanization, the question of *"Where to put the next billion people"*[30] calls an answer. The endless edge of cities has compelled the experts to set the global policy to secure and conserve hotspots biodiversity[31] and control soil consumption.

The experts sought to find some arbitrary agreements. For instance, in the dissertation *"A Global Assessment of the Links between Urbanization, Biodiversity, and Ecosystem Services"*,[32] the experts pointed out the chance of conservation *in* urban areas, stating that since many of the world's cities are in biodiversity-rich areas, cities have a vital role in conserving these critically threatened ecosystems (Elmqvist et al., 2013).

Moreover, Forman and Wu (2016) suggest an appealing proposal to concentrate the growth in four places on a global scale: "the outer suburbs; existing low-density sprawl areas just beyond the suburbs; satellite cities; and towns and villages within adjoining farmland". The overlapping of the following maps demonstrates the holistic effort of Forman and Wu to tackle global crises in line with the urbanization process. They identified habitable zones globally, where the coexistence of human and nature can be feasible without endangering biodiversity hotspots (Figure 4.2).

Due to the unavoidable process of urbanization, the ecosystems are changing from local scale to the global scales (Barnosky et al., 2012), and experts pursue "novel ecosystems" on a scale commensurate with the global crises (Hobbs et al., 2013).

Developing countries should pay for saving rainforests[33]; if tropical rain forests contribute to carbon storage and climate regulation of the globe; if the

---

[29]Title of the book Neil Brenner. (2014). *Implosions/Explosions: Towards a Study of Planetary Urbanization,* which reviewed the Lefebvre's hypothesis, stopping privileging cities in urban studies and start thinking critically about planetary urbanization.

[30]Tile of the article Forman and Wu (2016). *Where to put the next billion people.*

[31]Among steering committee on conservation of biodiversity, we can mention: CBD (Aichi targets for 2020), EEA (Natura 2000, LIFE), IUCN, UNEP, CITES, CMS, and GEF.

[32]See Convention on Biological Diversity (2012) *Cities and Biodiversity Outlook.* Available at: http://www.cbd.int/en/subnational/partners-and-initiatives/cbo.

[33]See Coalition for Rainforest Nations (CfRN), "Developing countries: pay us to save rainforests".

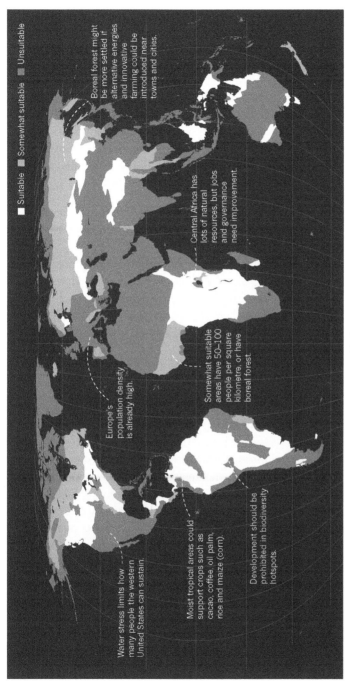

Habitable Zones

Places with warm and moist climates amenable to growing crops, such as grassy and forested lands in temperate and tropical regions, could in theory sustainably accommodate more people.

These include large areas of the Americas, central Africa and Asia as well as pockets of Oceania and Australia, but not populous or water-stressed regions or biodiversity hotspots.

Figure 4.2 Author's re-elaboration of Forman and Wu (2016) habitable zones. ©Nature 537, p. 608–611 (29 September 2016).

trees immigrate as well as human.[34] So, where are the boundaries between people and nature or cities and forests? Which is the scale of acting? In the era of the Anthropocene, we are all embedded. If the urbanization is boundless, so should be the equivalent forestation too. Weller and Hands (2014) state that:

> "If we accept that 'the city' is now a continuous system of global exploitation and not merely the morphology of various residential and commercial densities, then any discussion of the 'urban forest' means that we should also scale up our thinking and discuss the 'global city' and its relationship to the 'global forest'".

If the urbanization in China causes flooding di South Africa; If the wealthy countries are responsible for deforestation of tropical rainforests for the services they provide to the rest of the world.

If so, how can we proceed? If linking between designing and planning has already been challenging, how can we create linking between local, regional, and global scales? Shall we move through the boundaries of each scale? According to which criteria should we act? Top-down approach? Or a bottom-up approach? Is it just a matter of connectivity like green infrastructure? How about the geographical limits of cities?

4.5 "Puzzle Doer" of Forests and Cities on a Global Scale

The last sections stressed "holistic thinking", including many different types of relationships in a complex system. In order to create a "global forest" on a broad scale, we should proceed through the "forestation" on a finer scale.

Technically, a forest's quality and management in urbanized areas are different from a forest in a natural reserve. Furthermore, geographical limits (urban and non-urban), as mentioned, do not make sense when we talk about ecology[35] and urbanization (due to the expanded functional area). Therefore, to manage the relation between forestation and urbanization more effectively, it might be stressed on the characteristics of context rather than

---

[34] See Wohlleben, P. (2015). *The hidden life of trees*. Trans. Billinghurst, J. Vancouver Greystone Books.p.186/211. Wohlleben sustains that: "Trees are among the slowest-moving beings with which we share our world" (p.186) "Seeds, they are free. The moment they fall from the tree, the journey can begin" (p. 211).

[35] Euguene Odum (1993) specifies that the existence of two or more ecosystems and their interactions "results in new properties that do not exist in either of them".

the geographical edge. Each context poses different needs and wants. For instance, invasive species are rejected for creating a forest in most of the world, but in Shanghai, due to the saline soil, non-native plants assume an interest from an economic perspective; they require little care, and they are fast-growing. Further, they enhance the quality of soil (Zimmerli, 2010). Even though "sustainability is a global issue, its implementation is, to a high degree, local" (Schmeing, 2010).

The concept of landscape could explain the importance of context. Creating forests means creating landscapes, and "landscapes do not exist in isolation. Landscapes are nested within larger landscapes, that are nested within larger landscapes, and so on. In other words, each landscape has a context or regional setting, regardless of scale and how the landscape is defined. The landscape context may constrain processes operating within the landscape. The broad-scale processes act to constrain or influence finer-scale phenomena" (McGarigal, 2015). If we consider that the urban forests are imbedded in another context which is "city", and this whole is also embedded in another whole which is "global city", therefore, our initial forest is shifted, alternated, modified, and created according to the driving forces and actors of our initial city, and our initial city is forced and driven by the forces of global city.

This action and interaction of driving forces between city and landscape are explained by Richard Weller (2017); basing on the city form of Cedric Price (1982) "The City as an Egg", he poses the question: what is the form of the city today? Ironically, he writes, "perhaps the much vaunted compact, transit-oriented city would be an Italian frittata; instant cities such Dubai and Shanghai's Pudong might be egg-white soufflés; Andrés Duany's Seaside in Florida would surely be pavlova; and Detroit... a broken eggshell?" He explains that the missing point in Price's egg cities is lack of "reference to the various natural and cultural environments in which the eggs exist. Price's cities are all egg and no chicken". There is no matter what the form of the city is, but "where the egg comes from and where it is going". According to economic trajectories, political decisions, and social influences, statistics can foresee "where the egg is going".[36]

Franco Zagari (2013) connotes that the related dynamics between causes and effects of political, economic, and social issues are sometimes determined by factors very far from each other; like as famous chaos theory of "butterfly

---

[36]One example of this statistic can be seen in research work of "Urban Age". Available at: http://lsecities.net/ua/

effects" that the beat of its wings in the Amazon can produce who knows what effects, in who knows what time, and who knows where. The map of institutions and power is changed.[37]

If the cities result from chaotic causes and effects, it is essential to shift from urban geographical limits (urban, suburban, rural, city, and megacity) to the urbanization process (where population moves, why and how). Such deviation can generate a paradigm shift in "urban" forestation.

The trends and process of urbanization can be foreseen. Some cities are growing; others are shrinking. Some are formal and planned; others are informal. Some cities develop vertically, others horizontally. Each city interfaces with nature in a different way. The "essence" of each city is different. In keeping with the urbanization process, adequate forests can be adopted. For instance, the forests in shrinking cities assume distinct characteristics from the forests in high-density centres. The former city could have a "wild woodland" while the latter, due to lack of space and security of public space, probably could have a regular design park.

This proposal aims to operate by dialectical reasoning, understanding the process of urbanization and acting adequately in the field of forestation. It tends to give broad attention to the context, relationships, and background of forests rather than the benefits of forests *per se*. Creating forests are "ecologically" and "ethically"[38] correct, but paying attention to the global forces of the urban age is essential to delineate an accurate forest for each urbanized area.

Such an approach is neither planning nor designing but interacting. Planning is "what might be" (theory), and designing is "what it is" (output). Even though interacting rooted in both, it is flexible and based on challenging inputs. It is holistic thinking which "is teleological; it promotes integration and denies the quest for rigour when that rigour is achieved by partitioning our terrain. The defining characteristic of this tradition is its recognition of the epistemological necessity of comprehensiveness" (Verma, 1998).

Local cities construct a global city. Local cities broadly follow three processes of urbanization:

---

[37] Original Italian passage: "Il gioco fra cause e effetti politici, economici e sociali presenta dinamiche determinate da fattori a volte lontanissimi gli uni dagli altri, e il famoso battito delle ali di farfalla in Amazzonia può produrre chi sa quali effetti, in chi sa che tempi, chi sa dove. Cambia la mappa delle istituzioni e del potere"(Zagari, 2013, p. 47).

[38] Refers to "Land ethic" and the role of human as the steward of soils, waters, plants, and animals, broadening the boundaries of the community.

- Growth: in population and distension of cities—*Boom town.*[39]
- Shrinking: loss of inhabitants and breakup of urban structure (Beyer et al., 2006).
- Informal: the result of culturally driven individual and communal initiatives (Gouverneur, 2014).

It is not mechanistic thinking about "what" city and forest are; it is holistic thinking, which digs into "Why" is growing, shrinking, or self-constructing?[40] If we perceive the main reason, then we can propose the proper forest for each context; as Forman (2008) suggests, we can construct the big picture (global forest) by fitting small pieces (small forests):

> "I'd select from existing known good and bad urban patterns, and create others. I'm a puzzle doer. Puzzle pieces can be fit together based only on local compatibilities, but if you want faster progress, then also view the big picture for context. The spatial arrangement of important and compatible land uses would be my centrepiece for the sandbox, so everyone could see the big picture for the future. Spatial arrangement is a handle for wise planning a tangible flexible approach usable anywhere and at any scale, it offers promise for achieving a positive trajectory or even a successful result".

---

[39] Weller, R. (2009). *Boomtown 2050: Scenarios for a rapidly Growing City.* London: UWA Publishing.

[40] Refers to "Mechanistic thinking focuses on 'what' and holistic thinking digs into 'why'". Pearl Zhu. (2016). *Thinkinggair: 100 Game changing digital Mindsets to Compete for the future.* The book discusses the role of talented digital leaders and their holistic thinking.

References

Ahern, J. (1995). Greenways as a planning strategy. *Landscape and Urban Planning,*33, p. 131-155.

Amati, M., Taylor, L. (2010). From Green Belts to Green Infrastructure. Planning Practice & Research, 25(2), p. 143-155.

Amin, A., Thrift, N. (2002). *Cities: Reimagining the urban.* Cambridge; Polity press.

Barnosky, A. D., Hadly, E. A., Bascompte, J., Berlow, E. L., Brown, J. H., Fortelius, M., et al. (2012). Approaching a state shift in Earth's biosphere. *Nature*, 486(7 June). p. 52–58.

Bateson, G. (1979). *Mind and Nature: A Necessary Unity (Advances in Systems Theory, Complexity, and Human Sciences).* New York: Hampton Press, p. 13.

Beatley, T. (2016). *Hand book of Biophilic city planning and design.* Washington D.C.: Island press, p. 4.

Benedict, M. A. and McMahon, E.T. (2006). *Green Infrastructure: Linking Landscapes and Communities.* Washington, D.C.: Island Press.

Beyer, E., Hagemann, A., Rieniets, T., Oswalt, P. (2006). *Atlas of shrinking cities.* Stuttgart: Hatje Cantz Publishers, p. 26, 66–85, 104–107.

Clément, G. (1991). *Le Jardin en movement.* Paris: Pandora.

Daniels, R.E. (1988). The role of ecology in planning: some misconceptions. *Landscape Urban Planning* 15, p. 291-300.

Davidson, J., Wibberley, G. (1977). *Planning and the Rural Environment* Oxford: Pergamon Press.

Di Carlo, F. (2016). Paesaggi come auspici. In: Zagari, F. Di Carlo, F. (eds.). *Il paesaggio come sfida Il Progetto.* Melfi: Libria.

Dramstad, W.E., Olson, J.D., Forman, R.T. (1996). *Landscape ecology principals in Landscape architecture and Land-use planning.* Cambridge, MA, Washington D.C.: Harvard University Graduate school of Design and Island Press, p. 32.

Elmqvist, T., Parnell, S., Fragkias, M., Goodness, J., Guneralp, B., Scewenius, M., Sendstad, M., Macotullo, P.J., Seto, C.K., McDonald, R.I., Wilkinson, C.(eds.)(2013). *Urbanization, Biodiversity and Ecosystem Services: Challenges and Opportunities—A Global Assessment.* Dordrecht: Springer.

Fábos, J. G., & Ryan, R. L. (2004). International greenway planning: an introduction. *Landscape and urban planning*, 68(2), p. 143-146.

Firehock, K. (2010.) A Short History of the Term Green Infrastructure and Selected Literature. [online] *Charlottesville: Green Infrastructure Centre.* Available at: http://www.gicinc.org [Accessed 15 July 2017]

Forman, R.T.T. and Wu, J. (2016). Where to put the next billion people. *Macmillan Publishers Limited part of Springer Nature*, 537, p. 608-611.

Forman, R.T.T. (2015). Launching landscape ecology in America and learning from Europe. In: Barrett, G.W., Barrett T.L. Wu, J. (eds.). *History of Landscape Ecology in the United States*. New York: Springer, p. 13-30.

Forman, R.T.T. (2008). The urban region: natural systems in our place, our nourishment, our home range, our future. *Landscape Ecology* 23(3). p. 251-253.

Forman, R.T.T., Godron, M. (1986). *Landscape Ecology*. New York: John Wiley and Sons.

Gouverneur, D. (2014). *Planning and design for future informal settlements*. New York: Routledge.

Guinaudeau, C. (1987). *Planter aujourd'hui, batir demain: le préverdissement*. Paris: Collection Mission du paysage.

Hargreaves, G. (2007). Large Parks: A Designer's Perspective. In: Czerniak, J., Hargreaves, G. (eds.). *Large Parks*. New York: Princeton Architectural Press, p. 121-174.

Hobbs, R.J., Higgs, E.S., Hall, C. (2013). Novel Ecosystems: Intervening in the New Ecological World Order. Hoboken: Wiley-Blackwell.

Konijnendijk, C.C., Ricard, R.M., Kenney, A., Randrup T.B. (2006). Defining urban forestry_ A comparative perspective of North America. *Urban forestry & Urban Greening* 4, p. 93-103.

Konijnendijk, C.C. (2010). The Role of Forestry in the Development and Reform of Green Belts. Planning Practice & Research, 25(2), p. 241-254

Lawrence, H.W. (2008). *City Trees: A Historical Geography from the Renaissance through the Nineteenth Century*. Charlottesville: University of Virginia Press.

Lister, N.M. (2007). Sustainable large parks: Ecological design or designer ecology?. In: Czerniak, J., Hargreaves, G. (eds.). *Large parks*. New York: Princeton Architectural Press. p. 41.

Marshall, A. (2016). *Ecotopia 2121: A Vision for Our Future Green Utopia— in 100 Cities*. New York: Arcade Publishing.

McGarigal, K. (2015). Landscape Context. [online] *UMass Landscape Ecology, University of Massachusetts*. Available at: http://www.umass.edu/landeco/research/fragstats/documents/Conceptual%20Background/Landscape%20Context/Landscape%20Context.htm [Accessed 19August 2017]

McHarg, I. (1992). *Design with nature, 25th Anniversary ed*. New York: John Wiley and Sons. p. 29, 53, 101, 123, 124, 197.

Metta, A. (2008). *Paesaggi d'autore: il Novecento in 120 progetti*. Firenze: ALINEA EDITRICE. p. 98.

Moore, J. W. (ed.) (2016). *Anthropocene or Capitalocene? Nature, History, and the Crisis of Capitalism*. Oakland: PM Press/Kairos.

Muir, J. (1911). My First Summer in the Sierra. Boston: Houghton Mifflin.

Naveh, Z. (2001). Ten major premises for a holistic conception of multifunctional landscape. *Elsevier, Landscape and urban planning* 57. p. 269-284.

Scharmer, O., Käeufer, K. (2013). *Leading from the Emerging Future: From Ego-System to Eco-System Economies.* San Francisco: Berrett-Koehler Publishers.

Schmeing, S. (2010). Dagenham Docks, tree nursery as park. In: Ghiggi, D.(ed.). *Tree Nurseries Cultivating the Urban Jungle: Plant production Worldwide.* Baden: Lard Müller publisher.

Socco, C., Cavaliere, A., Guarini, S. (2007). L'infrastruttura verde urbana. [online] Osservatorio Città Sostenibili (OCS) Dipartimento Interateneo Territorio Politecnico e Università di Torino, Working paper P11. Available at: http://docplayer.it/10770473-Osservatorio-citta-sostenibili-dipartiment o-interateneo-territorio-politecnico-e-universita-di-torino-l-infrastruttura -verde-urbana.html [Accessed 16 August]

Steiner, F. (2014). Frontiers in urban ecological design and planning research. *Landscape and Urban Planning* 125, p. 304–311.

Steiner, F. (2011). Landscape ecological urbanism: Origins and trajectories. *Elsevier: Landscape and urban planning*, 100. p. 333-337.

Verma, N. (1998). *Similarities, Connections, and Systems.* Lanham: Lexington Books. p. 10-11.

Waldheim, C. (2017). Claiming Landscape as Architecture. In: Steiner, F. (ed.). *Cities and Nature.* Massachusetts: Lincoln Land Institute. p. 69-95.

Waldheim, C. (Ed.). (2006). *The Landscape Urbanism Reader.* New York: Princeton Architectural Press.

Weller, R. (2017). The City Is Not an Egg: Western Urbanization in Relation to Changing Conceptions of Nature. In: Steiner, F. (ed.). *Cities and Nature.* Massachusetts: Lincoln Land Institute. p. 31-50.

Weller, R., Hands, T. (2014). Building the Global Forest. [online] Scenario 04: Building the Urban Forest. Available at: https://scenariojournal.com/ar ticle/building-the-global-forest/ [Accessed 15 May]

Williams, K. (2000) Does intensifying cities make them more sustainable? In: K. Williams, M. Jenks & E. Burton, *Achieving Sustainable Urban Form.* London: Taylor & Francis.

Wilson E.O. (1984). *Biophilia.* Cambridge, MA: Harvard University Press, p. 31.

Zagari, F. (2013). *Sul paesaggio lettera aperta.* Melfi: Libria, p. 46-47

Zimmeli, C. (2010). Better city, Better Life. In: Ghiggi, D. (ed.). *Tree Nurseries Cultivating the Urban Jungle: Plant production Worldwide.* Baden: Lard Müller Publisher.

5

Urbanization and Forestation: Case Studies

Urbanization tendency and its relation to forests

> "God made the country and man made the town".
>
> William Cowper

This chapter, those mentioned above three main propositions of forestation based on the urbanization process, is validated and redefined through case studies.

The spatial and physical structures of forests in urban areas are the results of urbanization tendencies. The economic trajectory, historical, cultural, and social drivers generate the urban forms. Consequently, understanding where these urban forms are heading is crucial to propose an adequate forest type. The forestation scenarios discussed are all related to and driven by the scale, tendency, and forms of the urbanization process.

The selected case studies combine a patchwork of urban processes, with different scales, in a different location with contrasting problems to offer a broad and deep understanding of forestation in urban areas.

The tendency of urbanization has been broadly classified into three main types: developed cities, shrinking cities, self-constructed cities. Each category opts for complex cases, where the challenges of urbanization could be perceived as an opportunity for forestation.

This chapter tends to develop the scenarios for the upcoming century through an accurate interaction between the urbanization process and forestation. To tackle challenges that hover over humankind in the forthcoming century, we should plan the future of our planet, and it is necessary to approach the city holistically. The environmental disaster, land take, urbanization, and food shortage are all embedded in such a study.

The result of case studies indicates how deep comprehension of future transformation in urban areas can address the role of forestation for future challenges.

In the chapter *Developed Cities*, the case studies aim to demonstrate how the application of landscape urbanism, where forestation became the driver of territory transformation, has led to the great prototype of city-making.

In the urban age, where the scarcity of land is one of the main issues, the chapter *Shrinking Cities* highlights how vacant lands can provide an opportunity for food production, reducing soil consumption, acting as the Green Stormwater Infrastructure.

The chapter *Self-Constructed Cities* has drawn attention to forests' social and cultural role as part of collaborative initiatives. In a context where the lack of planning causes environmental and social disasters, the forestation could construct the "urban armature" for further urbanization.

The global forest could be created through the reunion of discussed scenarios in a global framework.

5.1 Developed Cities

Before the ground-breaking work of Indian economist Amartya Sen in the 1980s, the development was traditionally evaluated by economic criteria. Sen has drawn attention to human well-being as the new index for development (human development index).[1] A further assessment like quality of life (QOL) has outlined to measure the well-being of individuals and societies. Even though "there is no established convention for the designation of 'developed' and 'developing' countries" (UN, 2004),[2] this chapter refers to human well-being as the indicator of development. There is an implicit assumption that nature in cities promotes the health and well-being of inhabitants

Central Park, New York. (Source: author's photographic archive).

---

[1] According to Human Developments Reports of United Nations Development Programme (UNDP): The idea of human development focuses on criteria of fair opportunities for people to satisfy their desires.

[2] See https://unstats.un.org/unsd/mi/worldmillennium.htm

(Rohde and Kendle, 1994). According to UN Habitat (2013), the well-being of the population is associated with the environmental sustainability of cities.

The main concern in this chapter involves the process of urbanization and its integration into forestation in "developed" cities, where the prosperity and well-being of dwellers are preeminent. Expansion and densification are two central tendencies of urbanization for satisfying the demands of housing. In "developed" cities, such trends have been affiliated with green policy. However, the green policy has been mainly adopted as a subsequent plan; its holistic integration to urban plan has led to a "prosperous" model of city and nature. The spatial configuration adopted policy, benefits, and limitation of each model will be discussed in this section.

Densification, From Green Wedge to Pocket Forests: Copenhagen

Copenhagen was recognized as the European Green Capital in 2014. The root of its success returned to the late 1940s when the guiding plan of the city was published to limit the urban growth of Greater Copenhagen, called Finger Plan (*Egnsplan*) (Figure 5.1). This plan was inspired by Abercrombie's Greater London Plan (1944). The planners sought to organize the urban development on the base of an overall regional structure where urban development concentrated along five radial corridors (finger city) linked by railway system to the core of Copenhagen (the palm of hand) and the areas between the city fingers were separated by "green wedges" which has been kept exempt from development (Knowles, 2012; Andersen, 2008; The Finger Plan, 2015). The concept of green wedge policy is similar to Green Belt, but the green wedge is considered more flexible because it does not limit urban growth in an absolute way.

Although the Finger Plan faced some challenges and limitations,[3] it is considered a "flourishing" plan for two main reasons: its transit-oriented development (TOD) and the system of green wedges. TOD has been

---

[3]Upon the implementation of The Finger Plan in the late 1940s and 1950s, making the housing along the designated corridors was very desirable due to their proximity to rail lines and nature (Knowles, 2012). Nonetheless, with the development of private cars, during 1960s, the plan met some serious challenges, competition arose between public and private transportation, and traffic began to increase around the city. Later, in 1989, service facilities were constructed near railway corridors and this decreased the amount of commuter traffic into Copenhagen (Vuk, 2005, Næss et al., 2011).

codified later, in 1993 by Peter Calthorpe, to limit the sprawl of American cities, encouraging "people to live near transit services and decrease their dependence on driving".[4]

Figure 5.1   ©Danish Ministry of the Environment, Nature Agency—The Finger Plan, 2015.

---

[4]Peter Calthorpe. (1993). *The Next American Metropolis: Ecology, Community, and the American Dream.* New York: Princeton Architectural Press. See also http://www.tod.org/. TOD has been applied in Washington, Brisbane, Denver, Tokyo, and many other cities around the world (Curtis et al., 2009).

Other Nordic countries have adopted green wedges for the cities Helsinki and Stockholm[5] (Ståhle and Caballero, 2010). Recently, Fabiano Lemes de Oliveira (2017) has published the *Green Wedge Urbanism*, examining the global diffusion of the green wedge and its potential for sustainable cities.[6]

The original version of Finger Plan was purely an "urban" model; an "anthropocentric" design lacks any notion regarding ecology and landscape (Vejre et al., 2007). The railway systems constituted the network upon which the sprawl would have occurred, near the most appealing forests, lakes, and coastlines. Copenhagen's 1936 Green Network Plan focused on the aesthetic quality of the landscape for its conservation (Vejre et al., 2007).

Many tend to recognize the Finger Plan for its foresightedness; *de facto*, it is "more by accident than by design". TOD and the green wedge of Copenhagen have been introduced as a "sustainable" model of planning, while all of these were beyond the imagination of Bredsdorff and Rasmussen at that time (Vejre et al., 2007).

In the face of environmental awareness and ecological dispersal considerations in cites,[7] the void wedges (as the consequence of Finger Plan) have become part of the agenda for "sustainable" planning, and afforestation has implemented to create dense ecological corridor (Vejre et al., 2007). The inner green wedges have transformed from farmland to forests and gardens, creating new "rurban landscapes"[8] (Gullinck, 2004) to host the recreational amenities for the population. Several afforestation initiatives have launched by State Forest Services in the western fringe of Copenhagen.

During the last century, the old structure of Finger Plan has associated with new ecological functions. With the urban densification in Copenhagen, new plans have been implemented to increase the forests and parks in Copenhagen. For instance, the "pocket" parks (*Lommepark*) or small public urban green spaces (SPUGS) have actualized to guarantee an urban green space within 400 meters for at least 90% of the population in 2015 (Public Health Office Copenhagen, 2006). Small green parks are a compliment to large green wedge and fill the daily needs of being in contact with

---

[5]Since the green wedges allow the expansion in double directions, it has been adopted for the program "Building the city inwards" (1970) as sustainable regeneration of brownfield for growing population of Stockholm.

[6]One of the cases of green wedge that is discussed in this book is the *Raggi Verdi* of Milano.

[7]For instance, the idea of Greenway in USA, *Landscape Ecological Urbanism* by Forman and Godron, and *Design With Nature* by McHarg.

[8]*Rurban landscape* refers to the emergence of new landscape which is lack of identity of both urban and rural landscape.

nature. Furthermore, the Green Roofs Program (2010) enforces that "new buildings with flat roofs (less than 30-degree slope) must have green roofs".[9] The mentioned programs played essential roles to entitle Copenhagen as a biophilic city (Beatley, 2016). In fact, the whole city can be walked or cycled through its forests, parks, and gardens.

According to the i-Tree[10] Canopy (2015), 16.5% of the city is covered by trees,[11] which is not high compared to other European cities (e.g., Turin 16.2% and Geneva 21.4%[12]). From a regional scale, there is an unequal distribution of forests in Copenhagen. Most of the forests are placed in the north (e.g., haunting area of the monarch, Jaegersborg Deer Park) and west (afforestation in the 1970s of Vestskoven) of the green wedges. In contrast, the trees canopy in central areas is mainly developed, thanks to botanic gardens (Botanisk Have) and zoological gardens, which were founded for scientific purposes (Konijnendijk et al., 2005) and parks and gardens (like Vestre cemetery and Fælledparken).

Copenhagen is literately classified among the greenest cities[13] in Europe. This fact is not just relied on the amount of tree canopy of the city, but it is embedded in a holistic understanding of the urbanization process and its integration to pre-existence forests and lakes. Initially, it occurred unconsciously but later created the framework for the city, and more afforestation has implemented into the inner city. No matter how many trees and forests are planted in the city, the spatial configuration with the urbanization process, label Copenhagen as one of the greenest cities in Europe.

One of the most prominent examples of such avant-garde urbanization and its implementation to ecosystem services is the district of Ørestad. The outer expansion and the inner densification of Copenhagen have been compensated by Pocket forests (parks and garden) in dense areas and extensive forests in wedges (as a barrier to urbanization). The lessons learnt from Copenhagen ingrained in its holistic policy of afforestation in

---

[9]See: https://stateofgreen.com/en/profiles/city-of-copenhagen/solutions/green-roofs-in-copenhagen.

[10]i-Tree is a software suite from USDA Forest Service that provides urban forestry analysis, quantifying the structure of community trees and environmental services that trees provide.

[11]Available at: http://www.urbantreecover.org/location/copenhagen/

[12]See Treepedia: http://senseable.mit.edu/treepedia

[13]This fact relies also on other factors, like the water management, air quality, transport, waste and land management, and energy consumption, which all together define the sustainability of city.

accordance with the form of urbanization. In the future, planners who want to see the city correctly must keep themselves at "a necessary distance" to perceive urbanization and afforestation in a holistic framework.[14]

Expansion/Suburbanization and Large Forests: Paris

Since the early 19th century, many communities have developed around Paris as a place for employment and residence. A sizeable working-class population, who could not afford housing in Paris, was accommodated in the outskirts. During the last century, millions of people continued to migrate to the suburbs of Paris; the grands ensembles result from population explosion along the Parisian periphery (Teaford, 2011). The royal hunting grounds of Bois de Boulogne (west) and Bois de Vincennes (east) became incorporated into the expansion plan of the city; such integration is perfectly legible from the night map of Paris. Similarly, many other European royal hunting grounds have been embedded into the expanding network of the city. For instance, we can mention Hyde Park of London, Tiergarten in Berlin and Casa de Campo in Madrid (Berrizbeitia, 2007).

Further expansions in the north of Paris have led to Parc Départemental du Sausset in rural fringe areas. The presence of tree-lined streets and other large parks like Parc Georges-Valbon, Parc de la Villette in Paris' dense infrastructure has led to an equal distribution of trees in the urbanized context of Paris.

Bois de Boulogne and Bois de Vincennes preserve the history of their sites as the remnant of ancient forests and arise from the juxtaposition of roadways to connect the dispersed activities within the forests to its external context; they have been modified during the century and present as "palimpsest". By contrast, Parc Départemental du Sausset has been constructed on a large tract of open land to prevent encroaching development and restoring the region's rural landscape (Di Carlo, 2015; Hargreaves, 2007). Even though the origins and design processes for mentioned parks are absolutely varied, they have a familiar character of large parks[15] scale.

Today, as affirms George Hargreaves (2007), "it will be rare for a designer to come across the great park sites of yesterday" (like Bois de Boulogne

---

[14]Refers to "anyone who wants to see the earth properly must keep himself at a necessary distance from it" by Italo Calvino (1957), *The Baron in the Trees*. See also Fabio Di Carlo (2013). *Paesaggi di Calvino*. Melfi: Libria.

[15]According to *Large Parks* by Julia Czerniak and George Hargreaves (2007), the scale of Large has been defined as being greater than 500 acres (200 ha).

and Bois de Vincennes), and designers mainly "are making parks from the landscape that range from the artificial to landscape from blighted industrial areas to extremely toxic landscapes". If large forests (as parks) are not well-embedded into the context, they risk becoming "dispirited border vacuums" (Jacobs, 1961) and can "imply a break with context" (Czerniak, 2007).

The example of Parc Départemental du Sausset represents an accurate understanding of urban growth, whereas a large system of forests "preempt the city"[16] (Koolhaas, 1995) and sustain the urban development. Fabio Di Carlo (2015) explicitly points out that Corajoud reconsidered "the disciplinary position of landscape design as neither opposed to, nor derived from, other related disciplines, such as architecture and planning"; instead, landscape as the driver and "unique way of designing and perceiving territorial transformations, and thereby ultimately determined the existence of humanity on the planet". Such a view is what later landscape urbanism explicated as "the notion that landscape can be the primary determinant of urban transformation, replacing the previous idea of landscape as a balance to urban growth".

With the bourgeoning of the next billion urban dwellers in the future, suburban expansion will be unavoidable. In such scenarios, the examples like Parc Départemental du Sausset, as a pre-emptive process to urban growth (*préverdissement*), can play a vital role to transform the territory into a successful model of city making for forthcoming suburban. As Cranz (1982) writes, "Whatever is decided about the function of parks will largely derive from some vision of the city", large urban forests create "places for the imagination to extend new relationships and sets of possibility".

The expansion of Paris consciously or/and unconsciously contributes to the construction of large forests in its context. Large forests are associated with issues like ecology, ecosystem services, cultural meaning, and urbanization. If in "the past they succeed ingeniously in fulfilling these various equations", today, they should deal with "issues of stewardship, maintenance, cost, security, programming, and ad hoc populist politics" (Corner, 2007). However, "a dynamic master plan with flexible trajectories over time" (Hargreaves, 2007), which understands the demands of new development, can lead to newly constructed districts where resilience is durable.

---

[16]"The problem of bigness as a polemic in architecture by Koolhaas, has been adopted as a theorem in design of large parks, to generate a set of provocative questions like Does the large park "need the city", does it "represent the city", "pre-empt the city", or "is it the city?"" by Czerniak and Hargreaves (2007). *Large Parks.* p. 26.

Megaregions and Mega Forests: Northeast USA

Taking a glance at night satellite images of NASA (Figure 5.2) clarifies that the "urban" limits of most cities around the world have been blurred. The zone of their influence expands perceptively. The expansion of cities has generated a new scale of urbanization, called megaregions. Megaregions are economic actors that derive from Regionalism policy, where political and financial forces are concentrated in specific regions,[17] creating large networks of metropolitan areas where governance, infrastructure, and land use are planned at a regional urban scale (America 2050, 2012) (Figure 5.3).

Such an enormous physical scale of urbanization in which cities are just one form (Brenner, 2014) implies a new scale of afforestation. An analogous forest scale is required to compensate for the effect of such "giant monsters". Megaregions necessitate new comprehension of afforestation in an urbanized area. America 2050 is a national planning initiative, which focuses on the emergence of eleven megaregions in the United States, a developmental framework that conducts the necessary infrastructure for the upcoming population in the 21st century. A part of this plan regards the conservation of the landscape. The adopted landscape policy in Northeast megaregion[18] is the primary concern of this section.

By constructing a high-speed railway in Northeast megaregion, the planners tend to construct a TOD model, where the subsequent urbanization will occur, providing housing and services for additional 15 million residences (America 2050).

Megaregions reflect the interlocking economic systems and share natural resources and ecosystems, which require coordinated and broad-scale actions in a holistic framework (Regional Plan Association of America 2050, 2012). The Northeast Landscape Conservation list includes Green Infrastructure, biodiversity conservation, land consumption, climate change, and water management.

On a holistic scale, the main concerns that practitioners have adopted for forest management are biodiversity conservation (GI), watershed

---

[17] For instance, the European Union, as an example of Regionalism, derives from the efforts to deconceptualize European economic and political space, making national boundaries more permeable and fostering the functioning of the single European market (Deas and Lord, 2006).

[18] The Northeast is a densely developed economic powerhouse of United State, with approximately 19 million urban acres in the 13 states, which will upsurge to 22 million urban acres by 2040. The main cities of Northeast megaregions are Boston, New York, Philadelphia, Baltimore, and Washington D.C. (Regional Plan Association of America 2050, Northeast Landscape conservation, 2012).

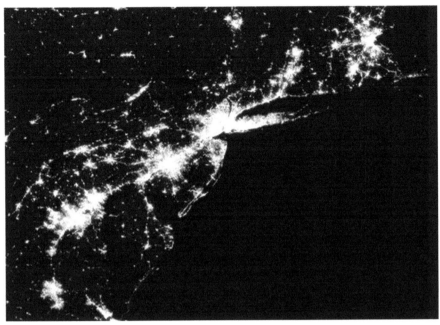

Figure 5.2   Composite image of Mid-Atlantic and Northeastern USA at night, 2016. ©NASA Earth Observatory images by Joshua Stevens, using Suomi NPP VIIRS data from Miguel Román, NASA's Goddard Space Flight Center.

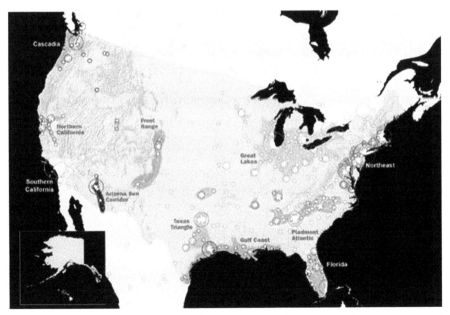

Figure 5.3   Author's re-elaboration of ©America 2050—Eleven emerging megaregions of USA.

management, and timber production.[19] On a smaller scale, a wide range of programs such as Green Infrastructure Stormwater (GIS) in Philadelphia,[20] Million Trees NYC initiative,[21] and water storm management in Brooklyn[22] have been conducted by public and private actors. Nonetheless, many practitioners affirmed a gap between small-scale projects and large landscape conservation programs.[23] The primary limits for coordinated collaborations are due to: "multistakeholder—they include public, private, and non-governmental actors, multipurpose—they address a mix of related issues" (water management, wood production, and biodiversity conservation); "multijurisdictional—the issues being addressed cut across political and jurisdictional boundaries" (Mc Kinney et al., 2010).

Regardless of the current complication between large and minor programs, the lesson learnt from Northeast Regional Plan is its holistic vision of urbanization and landscape conservation on a new scale which is "mega". The following maps (Figures 5.4 and 5.5) have been extracted from Landscape Conservation Report to present how a comprehensive vision of problems can adequately address the small issues on the local scale.

According to the urbanization framework, the first map suggests (Figure 5.4) expanding existing landscape initiatives to address distinctly urban challenges. The second map (Figure 5.5) highlights required open spaces in a densely populated area. Based on these maps, the "mega forest" of Northeast megaregion should be planted in the area within light purple to satisfy the accessibility to green areas.

To conclude, the new megaregion of the northeast operates across the traditional boundaries of cities. It is necessary to coordinate different actions based on an integrated framework to address issues at a large scale and fit them to the local scale solution (Porter and Wallis, 2002). The "mega forests" of Northeast megaregion should be planted "now" to compensate the footprint of additional 15 million upcoming people within 30 years.

---

[19]For instance, the New Jersey Highlands Water Protection and Planning Council is employing ecosystem services produced by protected forests to keep water quality safe for people living in northern New Jersey (Regional Plan Association and America 2050, 2012). See http://www.state.nj.us/njhighlands/

[20]See the next chapter, *Shrinking cities.*

[21]See http://www.milliontreesnyc.org/

[22]Brooklyn Greenway Initiative, Waterfront project of Brooklyn. See: http://www.brooklyn greenway.org/ and WEDesign for West Street in Greenpoint, Brooklyn, a resilience plan for storm water management. See: http://wedesign-nyc.com/portfolio/greenpoint-west-street-wa tershed-phases-i-ii/

[23]Low and Roman, interview, 21 June 2017; Kondo, interview, 27 June 2017.

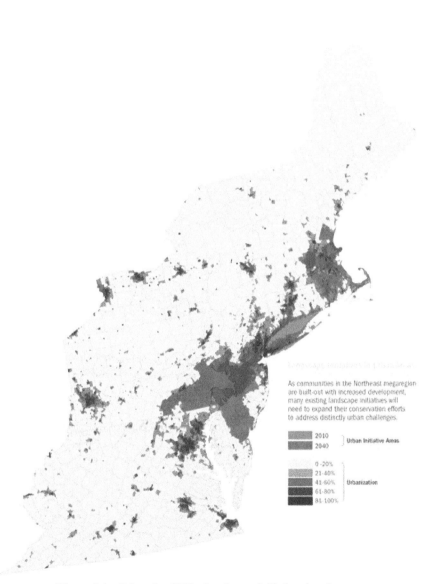

As communities in the Northeast megaregion are built-out with increased development, many existing landscape initiatives will need to expand their conservation efforts to address distinctly urban challenges.

2010	Urban Initiative Areas
2040	

0 -20%	
21-40%	
41-60%	Urbanization
61-80%	
81-100%	

Figure 5.4 ©America 2050—Landscape initiatives in urban areas.

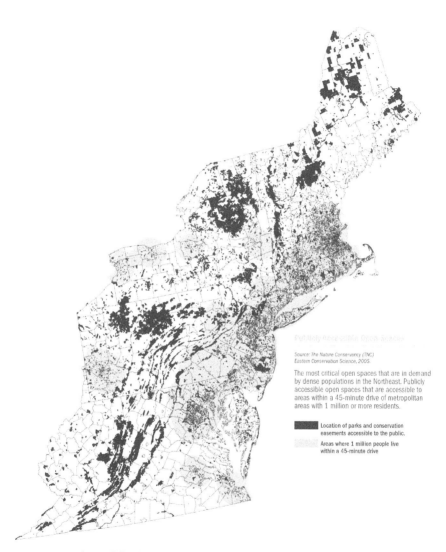

Source: The Nature Conservancy (TNC)
Eastern Conservation Science, 2005.

The most critical open spaces that are in demand
by dense populations in the Northeast. Publicly
accessible open spaces that are accessible to
areas within a 45-minute drive of metropolitan
areas with 1 million or more residents.

Location of parks and conservation
easements accessible to the public.

Areas where 1 million people live
within a 45-minute drive

Figure 5.5   ©America 2050—publicly accessible open space.

"Forests as the Driver of Territory Transformation"
Conclusion

Matthew Gandy (2014) puts forward the question: *Where does the city end?* In the past, different terms have emerged to describe the new forms of centrality in the peripheral areas: *edge city* (Garraeu, 1991), *technoburb* (Fisher, 1987) or *in-between cities* (Sieverts, 2002), or Anglo-American concept of "suburb" and French concept of *la banlieue*; later, other terms like "city-regions, urban regions, metropolitan regions, and mega cities represented a particular type of territory that was qualitatively specific and, thus, different from the putatively non-urban spaces that lay beyond their boundaries" (Schmid, 2016).

Regardless of the terms, as Gandy pinpoints, we should "focus on the urban process as a socioenvironmental dynamic that transcends the often-arbitrary distinction between the city and the "non-city".

The prominent examples of afforestation, like pocket forest and the green edge of Copenhagen, Large Parks of Paris and Mega forests of Northeast megaregion, descend from a profound understanding of the urban process. In the past, the form of Green wedge and Large Parks (Bois de Boulogne and Bois de Vincennes) came out "unconsciously", but Parc Départemental du Sausset was the result of deep comprehension of urban transformation, where the project of forest became the driver of the territory transformation (Di Carlo, 2015).

Suppose Northeast Megaregion of America tends to transform the territory through an adequate approach, it is necessary to focus on "mega" afforestation in a holistic framework, as the driver of territory transformation, and it should be done "now" to be able to compensate the footprint of following 15 million people within 30 years.

All around the globe, rapid urbanization is shifting towards novel forms and dimensions of the urbanized area. Extrapolating this shift to its consequences reveals a conclusion where the efficiency of forestation is approved only if the process of forestation exceeds urbanization. The mentioned case studies achieved such favourable outcome, just due to their capacity to be amended in line with the urbanization process.

References

Andersen, H.T. (2008). Copenhagen, Denmark: Urban Regeneration at Economic and Social Sustainability. In: Kidokoro, T., Harata, N., Subanu, L.P., Jessen, J., Motte, A., Seltzer, E.P. (eds.) *Sustainable City Regions: Space, Place and Governance*. Volume 7. Japan: Springer. p. 203-226.

Beatley, T. (2016). *Handbook of Biophilic City Planning & Design*. Washington: Island Press, p. 31.

Berrizbeitia, A. (2007). Re-placing Process. In: Czerniak, J., Hargreaves, G. (eds.). *Large Parks*. New York: Princeton Architectural Press, p.175-198.

Brenner, N. (ed.). (2014). *Implosions/Explosions: Towards a Study of Planetary Urbanization*. Berlin: Jovis Verlag.

Corner, J. (2007). Foreword. *Large Parks*. Ed. Czerniak, J., Hargreaves, G. New York: Princeton Architectural Press, p. 14.

Cranz, G. (1982). *The politics of park design*. Massachusetts: The MIT Press.

Curtis, C., Renne, J., Bertolini, L. (eds.) (2009). *Transit Oriented Development Making it Happen*. Farnham: Ashgate Publishing Limited.

Czerniak, J., (2007). Speculating on Size. In: Czerniak, J., Hargreaves, G. (eds.). *Large Parks*. New York: Princeton Architectural Press, p. 19-35.

Deas, I., Lord, A. (2006). From a New Regionalism to an Unusual Regionalism? The Emergence of Non-standard Regional Spaces and Lessons for the Territorial Reorganisation of the State. *Urban Studies* 43(10), p. 1847–1877.

Di Carlo, F. (2015). Michel Corajoud and Parc Départemental du Sausset. *Journal of Landscape Architecture*, 10(3). p. 68-77.

Fishman, R. (1987). *Bourgeois Utopias: The rise and fall of suburbia*. New York: Basic Books.

Gandy, M. (2014). Where does the city end?. In: Brenner, N. (ed.). *Implosions/Explosions: Towards a Study of Planetary Urbanization*. Berlin: Jovis Verlag, p. 86-89.

Garraeu, J. (1991). *Edge City: Life on the New Frontier*. New York: Doubleday.

Gullinck, H. (2004). Neo-rurality and multifunctional landscapes. In: Brabdt, J., Vejre, H. (eds.). *Multifunctional landscapes, Theory Values and History*. Southampton: Wit press, p. 3-31.

Hargreaves, G. (2007). Large Parks: A Designer's Perspective. In: Czerniak, J., Hargreaves, G. (eds.). *Large Parks*. New York: Princeton Architectural Press, p. 121-174.

Jacobs, J. (1961). *The death and life of great American cities.* New York: Random house, p. 257.

Knowles R.D. (2012). Transit Oriented Development in Copenhagen, Denmark: from the Finger Plan to Ørestad. *Journal of Transport Geography* 22, p. 251-261.

Konijnendijk, C.C., Nilsson, K., Randrup, T.B., Schipperijn, J. (Eds.) (2005). *Urban Forests and Trees.* Berlin: Springer.

Lemes de Oliveira, F. (2017). *Green wedge urbanism: history, theory and contemporary practice.* London; New York, NY: Bloomsbury Publishing.

Mc Kinney, M., Scarlett, L., Kemmis, D. (2010). *Large Landscape Conservation: A Strategic Framework for Policy and Action.* Cambridge: (MA) Lincoln Institute of Land Policy.

Næss P, Strand A, Næss T, Nicolaisen M. (2011). On their road to sustainability? The Challenge of sustainable mobility in urban planning and development in two Scandinavian capital regions. *Town Planning Review.* 82(3), p. 285-315.

Porter, D.R. and Wallis, A.D. (2002). *Exploring ad hoc regionalism.* Cambridge, MA: Lincoln Institute of Land Policy.

Public Health Office Copenhagen. (2006). Sunde Københavnere i alle aldre – Københavns Kommunes Sundhedspolitik 2006–10 (Healthy Copenhageners in All Ages – Health Policy of the Municipality of Copenhagen for 2006–2010). Available at: http://www.folkesundhed.k k.dk/Udgivelser/leadstoext/media/folkesundhedkbh/Politikker%20og%2 0strategier/sundhedspolitikkk.pdf.ashx [Accessed 15 June 2017]

Regional Plan Association of America 2050. (2012). Northeast Landscape conservation. [online] Available at: www.america2050.org/megaregions.html [Accessed 15 June 2017]

Rohde, C.L.E. and Kendle A.D. (1994). *Human well-being, natural landscapes and wildlife in urban area. A review.* Reading: English Nature Science.

Schmid, C. (2016). Planetary urbanization. In: Brenner, N. *Critique of Urbanization: Selected Essays.* Basel: Birkhauser.

Sieverts, T. (2002). *Cities without Cities: Between Place and World, Space, and Time, Town and country.* London: Routledge.

Ståhle, A., Caballero, L. (2010). Greening metropolitan growth: Integrating nature recreation, compactness and spaciousness in regional development planning. *International Journal of Urban Sustainable Development,* p. 64-84.

Teaford, J.C. (2011). Suburbia and Post-suburbia: A Brief History. In: Phelps, N.A., Wu, F. (eds.) *International Perspectives on Suburbanization A Post-Suburban World?* Hampshire: Palgrave Macmillan.

The Danish Nature Agency. (2015). The Finger Plan. Available at: www.naturstyrelsen.dk [Accessed 16 June 2017]

UN Habitat. (2013). *State of the World's Cities 2012/2013, Prosperity of Cities*. New York: Routledge.

Vejre, H., Primdahl, J., Brandt, J. (2007). The Copenhagen Finger Plan: Keeping a green space structure by a simple planning metaphor. In: B Pedroli, B., van Doorn, A., de Blust, G., Paracchini, M.L., Wascher, D., Bunce, F. (eds.). *Europe's living landscapes: Essays exploring our identity in the countryside.* Zeist: Koninklijke Nederlandse Natuurhistorische Vereniging, Stichting Ulitgeverij, p. 310-328.

Vuk, G., (2005). Transport impacts of the Copenhagen Metro. *Journal of Transport Geography*, 13(3), p. 223–233.

Websites:

www.America2050.org

www.rpa.org

5.2 Shrinking Cities

> "obsolete places in which only a few residual values seem to ménage to survive, despite their total disaffection from the activity of the city. They are, in short, external placet, strange placet left outside the city ... lacking any effective incorporation; they are interior islands voided of activity; they are forgotten, oversights and leftovers ...
>
> In short, these are places that are foreign to the urban system, mentally exterior in the physical interior of the city, appearing as its negative image as much in the sense of criticism as in that of possible alternative".

> de Solà Morales, I.[1]

Throughout history, many cities around the world have lost a significant number of inhabitants. Such phenomenon is described as "shrinking city" or "perforated cities", which illustrate the breakup of urban structure.

In *Atlas of Shrinking Cities* (Beyer et al., 2006), four main reasons have been mentioned for shrinkage:

- Destruction due to war, fire, natural disaster, and epidemics.
- Loss of resources, like water supply, oil, and mineral extraction.
- Shifting the locations of settlements or offshoring.
- Changes in economic, political, and demographic conditions.

According to McKinsey's report (2016), it is expected that from 2015 to 2025, the population decline in 17% of large cities in developed regions and 8% of all large cities across the world.[2]

Urban shrinkage is a "turning point" for architects and urban planners accustomed to planning the growth and must now deal with demolition and deconstruction (Dettmar, 2005). Many experts have focused their studies on such "in-between" terrains, coining diverse terms and functions like "non-place"[3] (Augé, 1995), "The Third Generation City",[4] "terrain

---

[1]de Solà Morales, I. (1996). Terrain Vague. *Quaderns* 212, Barcelona: Actar. p. 34-42.

[2]Available at:http://www.mckinsey.com/global-themes/urbanization/urban-world-meeting-the-demographic-challenge-in-cities Cities in Spain, Greece, and Russia are set to shrink fast. Most of the cities in Western Europe, United Sate, and Japan seem to shrink.

[3]Augé refers more to the buildings rather than open spaces (the perception of a place where human beings remain anonymous).

[4]See Casagrande, Taipei project in Taiwan. He proposes to step in nature in the ruins of industrial city, "recognizing its local knowledge and allows itself to be part of nature".

vague",[5] "The City of Nothing",[6] (Fabio Di Carlo, 2013), or "exaptation" (Gould and Vrba, 1982).[7]

Sociological problems like migration, ageing population, the decline in fertility (Kabisch et al., 2012) and new challenges for requalification of land (brownfield), and conservation of biodiversity are the central debates of shrinking cities (Haase, 2012; Haase, 2013). Nonetheless, it is believed that shrinkage provides opportunities to make cities greener (Schetke et al., 2010). The green management of shrinking cities has broadly studied in European countries, and such results can be adopted in other contexts, where due to the conflict (like African, middle east countries) or "hollowing out"[8] (like China) (Long et al., 2012), shrinkage is going to be the outcome. What is an adequate approach in such *terrain vague*?

Two main scenarios of green management can be considered; they both deal with the "placemaking"[9] strategy. The first is guided by nature and manifests the supremacy of natural evolution. The second is driven by a human, founded on human needs and wants, and the result is artificial.

Délaissé and *Friche*: German Experience

For many wastelands (like coal and steel industrial sites), the most ecologically, financially, and aesthetically appealing solution is to allow vegetation to grow on its own (Grosse-Bächle, 2005).

In his book *The World Without Us*,[10] American journalist Alan Weisman (2007) foresees the planet's future without human. He predicts that the forests

---

[5]Title of the movie Marcel Carné (1960) *"Terrain Vague"* depicts emblematic context of *banlieue* of Paris; later in 1996, de Solà Morales held to be true this term rather than any other term to explain the urban quality of abandoned areas.

[6]Di Carlo refers to *Invisible Cities* of celebrated Italian author Italo Calvino, talking about the consequences of modern urbanization and its residues.

[7]Exaptation is used by biologist to express the evolution and adaption process of organism; architectural use of this term represents the potential of vacant land.

[8]Referring to rural communities where villagers move to new cities for jobs.

[9]Placemaking provides new perception of a place and contributes to the well-being of communities. See also Fleming, L. (2007). *The Art of Placemaking: Interpreting Community Through Public Art and Urban Design.* London and New York: Merrell Publishers.

[10]This book is an expansion of the article "Earth without people" by Weisman, A. (2005) in *Discover Magazine,* 26(2). p. 60-65. He applies more ecological view to *Earth Abides* of George R. Stewart (1949), which tells the history of the death and rebirth of civilization.

would substitute the residential areas after 500 years. Even though it seems speculative fiction, it represents the power of nature to vanish the human traces in abandoned lands.

Gilles Clément (2004), a French gardener, affirms the power of nature in an "unattended" area, where the diversity turns up by its evolution. The *délaissé* (neglected land) and *friche* (wasteland) are the spaces that give birth to the *Third Landscape*—"an undetermined fragment of the Planetary Garden - designates the sum of the space left over by man to landscape evolution - to nature alone".

The wild woodlands as a significant natural evolution of forests are widely debated in abandoned contexts. The contributions from the conference *Wild Forests in the City–Post-Industrial Urban Landscapes of Tomorrow*,[11] which has been published in the book *Wild Urban Woodlands, New Perspectives for Urban Forestry*, draw attentions to the role of wild woodlands in altered urban-industrial areas: "[...] Wild urban woodlands resulting from natural succession on man-made sites have created a new component in the urban forest mix, whose significance will grow in areas that are subjected to great structural transformation. These include many former industrial areas, but also, more generally, 'shrinking cities'" (Kowarik et al., 2005).

Many European post-industrial countries have adopted wild woodlands as a strategy to enhance biodiversity and ecosystem services (Kowarik et al., 2005) in their shrinking cities. The most prominent examples are in Germany: abandoned rail yards like Natur-Park Südgelände—in Berlin designed by Odius landscape group, Park am Gleisdreieck by atelier LOIDL, Industrial area of the Ruhr projected by Latz and Partner, and lignite mining areas of eastern Germany like Leipzig and Lusatia that are transformed into navigable lakes surrounded by forests designed by Stefan Giers.

Regardless of the origin of shrinking (ex-industrial area or largely abandoned cities), the main argument is that "the newly created open spaces must give a new organizational pattern to the whole" (Dettmar, 2005). Wild woodlands fade away ruined structure and form a new edge of space. Such space can be an open landscape for recreational purposes. One of the recent recommended activities is *Forest Bathing (shinrin-yoku)*,[12] which potentially relieves disease and is considered an alternative health-promotion approach (Yu et al., 2017; Tsunetsugu et al., 2010). Although opening a site to the

---

[11] It has been organized by the Institute of Ecology at the Technical University Berlin in cooperation with the *Projekt Industriewald Ruhrgebiet* and took place from 16–18 October 2003 in Dortmund, Germany.

[12] "Shinrin-yoku" is "Therapeutic Effects of Forests" which received attention in Japan.

public endangers biodiversity, the concept of zoned spaces could apply to control partial access to the site with higher biodiversity (Davies, 2005).

Even though wild woodlands are widely proposed in industrial areas, Ryohei Ono (2005) considered the cultural aspect of this "new" landscape. He mentions that in Japan, the forests are mainly located on the hill or mountain and have roots in the religious belief of Shinto.

Diagrams of industrial development (above) and the different narratives in the post-industrial city (below).

This configuration has created a "cultural image" of the hill, analogous to the ancient religious belief and had a grand narrative. The brownfields of industrial development are located near the sea; planting forests in the post-industrial site will be against the bygone memory of the landscape. "Even if we could build forests in the cities using ecological technology without any cultural context, we cannot be sure that the forests would be accepted by people and built into the new narratives of the cities" (Figure 5.6).

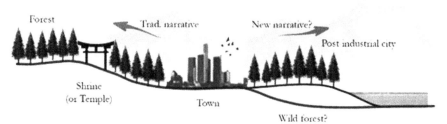

Figure 5.6   Author's re-elaboration of ©Ryohei Ono (2005).

This example has been mentioned to specify that not always incorporating such woodlands into the green development are feasible, and sometimes few conditions are required.

To warp up, ecologically talking wild woodlands are an astute method for shrinking land and can express the desire of Gilles Clément: *Je voudrais montrer la diversité extrême de ce qui existe sur la planète.* They assume the role "as the genetic reservoir of the planet, the space of the future".

From "Food Deserts" to Trees Canopy: Detroit

> "If you lived next to a vacant house and now all of a sudden you live next to a forest, you are in better shape".

> Joseph van Dyk
> Gary's director of planning and redevelopment

Like other shrinking cities of the North American Rust Belt, Detroit has suffered the deindustrialization severely after 1950. Racial conflict and mobility of car-driven society caused the inner city to shrink while the suburbs continued to sprawl.[13] The inner city populated by African American impoverished, while the boroughs of white middle-class prospered (Beyer et al., 2006).

The unequal distribution of services in the perforated city of Detroit influenced the food access issues. Disregarded by supermarkets, the poor dwellers of inner areas have to go far to shop in a food store (Gallagher, 2007). Detroit was coined as *food deserts* (Zenk et al., 2005). The initiatives by residents go against ongoing depreciation and dereliction of vacant lots. They convey the urban agriculture movement into the vacant lots; Detroit became the prominent example in the United States to manage abandoned lands (Morgan, 2015). Vacant lots contribute to "green" land use like reforestation, installation of parks, community gardens, and food production. The trees plantation plays the role of ecosystem services and contributes to the economy of the city. The city has engaged Recovery Park,[14] aiming to put back the vacant lots into productive use, revitalizing neighbourhoods, improving the environment through sustainable management.

Nonetheless, Paddeu (2017) explains that the economic benefits come first compared to equity and ecology; urban agriculture is the last

---

[13] Such a phenomenon is called "doughnut effect": an appealing suburban area with post-industrial wastelands in the middle.

[14] See http://www.recoverypark.org/

strategy in shrinkage city. The legal access to the vacant lands becomes bothersome. Farming on brownfields with contaminated soils, chemicals, or Genetically Modified Organisms discredit the environmental benefits of urban agriculture. Furthermore, some experts believe that urban agriculture has been adopted to evict African Americans from a planned context (Rhodes and Russo, 2013).

However, the "greening" issues found a haven in all shrinking cities of the Rust Belt (Paddeu, 2017). The idea of "Creative Shrinkage" (The New York Times Magazine's, 2006)[15] "Smart decline" (Hollander and Németh, 2011) draws the attentions of politician and planners in depopulated Rust Belt cities to transform vacant lots into pocket parks and forests (Schwarz and Rugare, 2008). Much research has focused on Cleveland, Youngstown, and Gary to re-conceive cities, seeking nontraditional approaches for restoring these *terrains vagues* (Aeppel, 2007; Bird, 2016).

The Rust Belt shrinking cities are located in a "habitable zone" defined by Forman and Wu (2016). If the fragmented lands of the Rust Belt might be planned in a holistic vision, the megaregional infrastructure of forests (Schilling and Logan, 2008), farms, and parks would become adequate for the upcoming *next billion people* without further land consumption for housing.[16]

Illicit Activities and GSI: Philadelphia

All around the world, apprehension exists about potential crime associated with wild and dense vegetation in green areas. For instance, Golden Gate Park was set up by homeless and sex workers, and just after the vast fire by arsonists, the camps were ceased by the mayor. Prostitution in Bois de Boulogne and Casa de Campo and the presence of racist in Landschaftpark Duisburg-Nord, all depict the social unrest in overgrown vegetated areas (Beardsley, 2007). Illustrated examples clarify that not always wild woodlands are the most appropriate answer, mainly when vacant lots are near other inhabited areas in a shrinkage city.

It has been estimated 40,000 vacant lots in the post-industrial city of Philadelphia.[17] Properties values in vacant lots plunge while crime rates

---

[15] Available at: http://www.nytimes.com/2006/12/10/magazine/10section1B.t3.html?mc ubz=0

[16] See the chapter *Holism* of this book; Forman R.T.T. and Wu, J. (2016). Where to put the next billion people. *Macmillan Publishers Limited part of Springer Nature,* 537, p. 608-611.

[17] According to Redevelopment Authority of the City of Philadelphia and Philadelphia Association of Community Development Corporations 2010.

(overgrown vegetated lots have created an excellent place for hiding drugs and guns) and vandalism (trash and fear) increase (Kondo, 2015; Garvin et al., 2013). Tax delinquents are applied to the citizens. These issues heighten with an irreversible vicious cycle (Huang, 2014).

The city of Philadelphia has adopted innovative and untested programs to abate vandalism. One program regards the vacant-lot greening, single trees are planted in lots, and weeds are removed to increase the surveillance of each lots; such a method reduced gun assaults and vandalism and improved the health outcomes (Branas et al., 2011). Another program is the construction of Green Stormwater Infrastructure (GSI) (such as forests, rain gardens, bioswales, and wetlands) in vacant lots to capture the stormwater runoff[18] (Low, 2014, Kondo et al., 2015; FAO, 2016).

Nonetheless, most vacant lots are private properties, and it can be a challenge for government agencies to spend money on forest implementation on private lands. Two methods have been adopted to provide a solution: in Philadelphia, a separate stormwater fee is charged to the properties based on impervious surfaces. A cost-saving has been offered to owners who install green infrastructure in their lots (Low, 2014). Furthermore, Philadelphia Land Bank (2015) opened a new mandate with the strategy "to quicken the pace of returning vacant and tax-delinquent property to productive use by centralizing ownership of these parcels". The lots which are not desirable to for-profit real estate "work best as small community gardens, side yards or open space until the neighbourhood recovers" (Smith, 2017).

To summarize, trees are planted in vacant lots of shrinkage city of Philadelphia to tackle illicit activities and stormwater runoff. Even though some are not the most appealing forests, they demonstrate "The Power of a Green Lot".[19] Philadelphia is a rigorous example of a holistic approach in which forestation is associated with other significant social issues in a shrinkage city.[20]

---

[18]See Philadelphia Water Department 2011 *Green City, Clean Waters*. Runoff is a main issue in Philadelphia; since the city is located near major water ways at downstream points in water shed, it receives end of stormwater runoff produced by neighbouring communities. Even though there are other sewage treatment plant, during high stormwater, these systems exceed their capacity.

[19]The online journal *Hidden City of Philadelphia* entitled this project as *The Power of a Green Lot*.

[20]The contribution of this project to health and safety of inhabitants becomes the topic in Medicine faculty of University of Pennsylvanian. See *MoreGreen, Less Crime: Rehabilitating Vacant Lots Improves Urban Health and Safety, Penn Study Finds.*

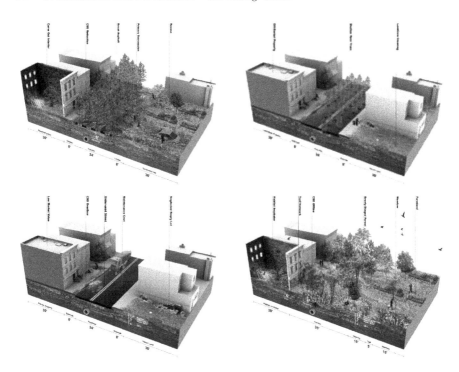

"©Chieh Huang (2014)
Urban Regeneration: Foresting Vacancy in Philadelphia.[21]

From Non-place to Place Making
Conclusion

To conclude, *place making* in shrinking cities is deeply connected to the size, location, and usefulness of lots (Rink, 2005). The wild woodlands are ecologically, financially, and aesthetically the most acceptable answer. Although they are the most preeminent solution in sizable land (like ex-industrial areas), woodlands in small-scale lots (among residential areas) can provoke vandalism (Beardsley, 2007; Davies, 2005). Small lots can be associated with other social amenities like GSI (like rain gardens), urban farms, orchards or parks. Surveillance and legislation are required to ensure the environmental benefits of these lots.

Available at: https://www.pennmedicine.org/news/news-releases/2011/november/more-green-less-crime-rehabil

[21] For details of this proposal, see: https://scenariojournal.com/article/urban-regeneration/

Shrinkage provides an opportunity for incrementing green space, and, with an adequate policy of land reclamation, it would decrease soil consumption. Such cities can increase the valuable land to accommodate the upcoming next billion people of our planet.

References

Aeppel, T. (2007). Youngstown thinks small as its population declines. [online]*Wall Street Journal*. Available at: https://www.wsj.com/articles /SB117813481105289837 [Accessed 26 August 2017]

Augé, M. (1995). *Non-places: Introduction to an Anthropology of Supermodernity*. London: Verso.

Beardsley, J. (2007). Conflict and erosion: The contemporary public life of Large parks. In: Czerniak, J., Hargreaves G. *Large Parks*. New York: Princeton Architectural Press.

Beyer, E., Hagemann, A., Rieniets, T. , Oswalt, P. (2006). *Atlas of shrinking cities*. Stuttgart: Hatje Cantz Publishers, p. 26, 66–85, 104–107.

Bird, W. (2016) Hard-Pressed Rust Belt Cities Go Green to Aid Urban Revival. [online]*Yale Environment 360, Yale School of forestry and Environmental Studies*. Available at: http://e360.yale.edu/features/g reening_rust_belt_cities_detroit_gary_indiana [Accessed 24 August 2017]

Branas, C.C., Cheney, R.A, MacDonald, J.M., Tam, V.W., Jackson, T.D., Ten Have, T.R. (2011). A Difference-in-Differences Analysis of Health, Safety, and Greening Vacant Urban Space. *American Journal of Epidemiol,* 174, p. 1296–1306.

Clément, G. (2004). *Manifeste du Triers paysage*. Paris: Éditions Sujet/Objet.

Davies, C. (2005). People Working for Nature in the Urban Forest. In: Kowarik, I. & Körner, S. (Eds.). *Wild urban woodlands: New perspectives for urban forestry*. Berlin: Springer.

Dettmar, J. (2005). Forests for Shrinking Cities? The Project "Industrial Forests of the Ruhr". In: Kowarik, I., Körner, S.(eds.). *Urban Wild Woodlands*. Berlin: Springer. p. 263–276.

FAO. (2016). *Guidelines on urban and per-urban forestry*. By Salbitano F., Borelli, S., Conigliaro, M., Chen, Y. FAO Forestry Paper No. 178. Rome.

Gallagher, M. (2007). Examining the Impact of Food Deserts on Public Health in Detroit. [online] Mari Gallagher Research &Consulting Group.

Available at: http://www.uconnruddcenter.org/resources/upload/docs/what /policy/DetroitFoodDesertReport.pdf [Accessed 26 August 2017]

Garvin, E., Branas, C., Keddem, S., Sellman, J., Cannuscio, C. (2013). More Than Just an Eyesore: Local Insights and Solutions on Vacant Land and Urban Health. *Journal Urban Health*, 90.

Gould, S., Vrba, E. (1982). Exaptation—a Missing Term in the Science of Form. *Paleobiology*, 8(1), 4-15.

Grosse-Bächle, L. (2005). Strategies between Intervening and Leaving Room. In: Kowarik, I., Körner, S.(eds.). *Wild urban woodlands: New perspectives for urban forestry*. Berlin: Springer. p. 231–246.

Haase, D. (2013). Shrinking Cities, Biodiversity and Ecosystem Services. In: Elmqvist, T., Fragkias, M., Goodness, J., Güneralp, B., Marcotullio, P.J., McDonald, R.I., Parnell, S., Schewenius, M., Sendstad, M., Seto, K.C., Wilkinson, C.(eds.). *Urbanization, Biodiversity and Ecosystem Services: Challenges and Opportunities. A Global Assessment.* NY, London: Springer.

Haase, D., Kabisch, N., Haase, A., Kabisch, S., Rink, D. (2012). Actors and factors in land use simulation – The challenge of urban shrinkage. *Environmental Modelling and Software, 35.* p. 92–103.

Hollander, J., Németh, J. (2011). The bounds of smart decline: a foundational theory for planning shrinking cities. *Housing Policy Debate*, 21.p. 349-67.

Huang, C. (2014). Urban Regeneration: Foresting Vacancy in Philadelphia. [online] Scenario 04: Building the Urban Forest. Available at: https://scen ariojournal.com/article/urban-regeneration/ [Accessed 20 August 2017]

Kabisch, N., Haase, D., & Haase, A. (2012). Urban population development in Europe, 1991– 2008: The examples of Poland and the UK. *International Journal of Urban and Regional Research*, 36(6). p. 1326–1348.

Kondo, M.C, Low, S.C., Henning, J., Branas, C.C. (2015). The Impact of Green Stormwater Infrastructure Installation on Surrounding Health and Safety. *American Journal of Public Health*, 105(3). p. 114-121

Kowarik, I. & Körner, S. (2005) (Eds.). *Wild urban woodlands: New perspectives for urban forestry.* Berlin: Springer.

Long, H., Li, Y., Liu, Y., Woods, M., Zou, J. (2012). Accelerated restructuring in rural China by 'increasing vs. decreasing balance' land-use policy for dealing with hollowed villages. *Land Use Policy*, 29(1). p. 11–22.

Low, S.C. (2014). Cumulative Effects: Managing Natural Resources for Resilience in the Urban Context. USDA Forest Service RMRS-P-71. p. 393-401.

Morgan, K. (2015). Nourishing the city: the rise of the urban food question in the Global North. *Urban Studies*, 52(8). p.1379-1394.

Ono R. (2005). Approaches for Developing Urban Forests from the Cultural Context of Landscapes in Japan. In: Kowarik I., Körner S. (Eds.) *Wild Urban Woodlands*. Berlin: Springer.

Paddeu, F. (2017). Legalising urban agriculture in Detroit: a contested way of planning for decline. *The Town Planning Review*88(1), Liverpool University Press. p. 109-129.

Rhodes, J. and Russo, J. (2013). Shrinking 'smart'?: Urban redevelopment and shrinkage in Youngstown, Ohio. *Urban Geography*, 34. p. 305-326.

Rink, D. (2005). Surrogate Nature or Wilderness?. In: Kowarik, I. & Körner, S. (2005) (Eds.), *Wild urban woodlands: New perspectives for urban forestry.* Berlin: Springer.

Schetke, S., Haase, D., Breuste, J. (2010). Green space functionality under conditions of uneven urban land use development. *Land Use Science, 5*(2). p. 143–158.

Schilling, J., Logan, J. (2008). Greening the Rust Belt: A Green Infrastructure Model for Right Sizing America's Shrinking Cities. *Journal of the American Planning Association* 74(4). p. 451-466.

Schwarz, T., Rugare, S. (eds.) (2008). *Cities Growing Smaller*. Urban Infill, Volume 1. Cleveland Urban Design Collaborative Kent State University.

Smith, S. (2017). Philadelphia has a new plan for its 43,000 vacant lots. [Online] *Next City*. Available at: https://nextcity.org/daily/entry/phila delphia-land-bank-2017-vacant-lots [Accessed 24 August 2017]

Tsunetsugu, Y., Bum-Jin Park, B.J. Miyazak, Y.(2010).Trends in research related to "Shinrin-yoku" (taking in the forest atmosphere or forest bathing) in Japan. *Springer,* Environmental Health Preventive Med*icine*, 15(1), p. 27–37.

Yu, C.P., Lin, C.M., Tsai, M.J., Tsai, Y.C., Chen, C.Y. (2017). Effects of Short Forest Bathing Program on Autonomic Nervous System Activity and Mood States in Middle-Aged and Elderly Individuals. *International Journal of Environmental Research and Public Health*,14(8): 897.

Zenk, S., Schulz, A., Israel, B., James, S., Bao, S., Wilson, M. (2005). Neighborhood racial composition, neighborhood poverty, and the spatial accessibility of supermarkets in Metropolitan Detroit. *American Journal of Public Health*, 95. p. 66-67.

Websites:
Vacant lot Philadelphia:
http://www.groundedinphilly.org/

Vacant lot Detroit:
https://www.motorcitymapping.org/#t=overview&s=detroit&f=all&c=board
ing

5.3 Self-Constructed Cities[1]

> "Close under the Abbey of Westminster there lie concealed
> labyrinths of lanes and potty and alleys and slums, nests of
> ignorance, vice, depravity, and crime, as well as of squalor,
> wretchedness, and disease; whose atmosphere is typhus, whose
> ventilation is cholera; in which swarms of huge and almost
> countless population, nominally at least, Catholic; haunts of filth,
> which no sewage committee can reach – dark corners, which no
> lighting board can brighten".

Wilfrid Philip Ward, 1897[2]

Self-constructed cities are "the result of culturally driven individual and communal initiatives" (Gouverneur, 2014). Soon, such a phenomenon will be the main form of urbanization in many parts of the world (Zárate, 2016). Although shelter is considered a fundamental human right, one-third of the global urban population suffers from inadequate living conditions (Habitat, 2013). Stakeholders and planners often deny dwellers of these areas' natural amenities, infrastructure, and services. They mainly suffer from lack of essential services, health hazards, insecure tenure, overcrowding, poverty, social segregation, and ineffective governance (UN-Habitat, 2003a; Gouverneur, 2014; Zárate, 2016). Even though they may locate within the administrative boundary of a town or city, "officially, they do not exist" (UN-Habitat, 2003b).

Self-constructed cities mainly are in Latin America, Sub-Saharan Africa, and Southeast Asia (UN-Habitat, 2013) and represent a significant portion of

---

[1]Other terms used to refer to such impoverished neighborhoods: Squatter settlements or Shanty town or Shacks (Daniel Carter Beard, *Shelters, Shacks, And Shanties,* 1914), Favelas (refers to *Favela,* hill outside Rio de Janeiro).Villas miseria (Bernardo Verbitsky, *Villa Miseria también es América,* 1957); Slum (Wilfrid Philip Ward, *The Life and Times of Cardinal Wiseman,* 1897); Bidonvilles, rookery, gecekondu, skid row, barrio, ghetto, taudis, bandas de miseria, barrio marginal, morro, loteamento, barraca, musseque, tugurio, solares, mudun safi, karyan, medina achouaia, brarek, ishash, galoos, tanake, baladi, hrushebi, chalis, katras, zopadpattis, bustee, estero, looban, dagatan, umjondolo, watta, udukku, and chereka bete (UN-Habitat, 2003, page 30).

[2]This quotation was widely cited in the national press to popularize the word slum to describe bad housing. Cited by Dyos, H.J., Cannadine, D., Reeder, D. (1982). *Exploring the Urban Past: Essays in Urban History.* Cambridge: Cambridge University Press. p. 240 and Wohl, A.S. (2002). *The Eternal Slum: Housing and Social Policy in Victorian London.* New Jersey: Transaction Publishers. p. 5.

the world's poor (World Bank, 2008). These communities are mainly located near endangered bioregions on the Earth. Due to the poverty, many of these communities are forest-dependent, with an additional 400–500 million people estimated to be directly dependent (White and Martin, 2002). The inhabitants of these settlements use the forest for hunting animals and harvest fruit, seeds, and woods (Vanda, 1993; IBDF, 1981; Oberndorf et al., 2007; FAO, 2003, FAO, 1993).

The continual growth of self-constructed cities will inevitably aggravate the human impact on our planet. Such settlements are encroached onto environmentally sensitive areas, populating dangerous territories with geological and physiological limitations. If self-constructed cities are appropriately guided, they can reduce poverty, environmental degradation, and disaster vulnerability (Parker et al., 1995).

This chapter aims to evaluate the role of urban forestry in the self-constructed cities of Cartagena in Colombia and San José de Chamanga in Ecuador. Each of these cases tackles different issues profoundly connected to rapid urbanization and its consequences.

Although the proposals emerge from academic studies, they illustrate the diversity of urban forestry conditions in informal settlements. The mentioned case studies were conducted in the Landscape Architecture Department at the University of Pennsylvania in 2017 and supervised by David Gouverneur and Maria Altagracia Villalobos.

The goals of such an approach encompass two main issues: identification of an acceptable method for urbanization and poverty reduction. The former has been applied through the plantation of the Dry forest in Cartagena as "Informal Armature"[3] (Gouverneur, 2014) to define the direction and location of forthcoming settlements; this novel strategy of planning allows citizens to be able to shape their environment in an adequate manner, where nature is not considered an impediment or afterthought but integrated into the process of self-organization. The latter has been examined through the role of Mangroves as an economic resource in poor settlements of San José de Chamanga in Ecuador.

Albeit it is stated that self-constructed cities devastate natural resources, David Gouverneur (2014) elucidates that in comparison with large cities of

---

[3]The "Informal Armature" is proposed by Gouverneur as a novel strategy that takes into consideration the vitality of informality and attempts to couple flexible design with management issues and supports the efficient use of resources for the inhabitants of informal settlements, who will have the capacity to shape their own habitat. It is not based on rigid model of planning; albeit, it provides a structure for informal developments.

the developing world, "[...]the informal city is compact, pedestrian-friendly, and socially cohesive. It can incorporate mixed uses at a neighbourhood scale. It consumes little energy and produces relatively low quantities of solid waste[...]These positive attributes facilitate the task of creating a sustainable future for cities [...]". Based on this statement, the case studies examine the role of urban forestation as an approach towards the sustainability of informal contexts.

Dry Forest (Dryfo) Urbanism: Cartagena[4]

Since colonial times, the thriving ports of the Spanish Colony were located on the Caribbean coast (Moya Pons, 2007). Cartagena was the main port for trade and exports of products from countries located in the Pacific basin of South America (Grahn, 1996). Today, 72% of the total population of the Caribbean coast is concentrated in commercial and touristic cities of Cartagena, Barranquilla, and Santa Marta (DANE, 2013).

The Caribbean coast is a beloved destination for national and foreign visitors for "Sun, Sea and Sand", Cartagena, San Andres and Santa Marta are the most visited cities (Rangel-Buitrago et al., 2015), and tourism represents significant economic activity in these areas.[5]

These rapidly urbanizing centres are contiguous to diverse systems of protected and endangered natural areas such as mangroves and dry forest, and tropical rainforest (Barreda-Bautista et al., 2011; Rangel-Buitrago et al., 2015).

Tourism activities and enlargement of existing coastal towns and villages aggravated the coastal erosion and destruction of mangroves (Rangel-Buitrago et al., 2015) with associated impacts on landscape scenery (Rangel-Buitrago et al., 2013). It is expected that by 2050, if the growth trends continue, Cartagena and Barranquilla will become a unified linear city along the coast (Figure 5.7). Historically, the rapid urbanization process was

---

[4]All rights reserve to Ishaan Kumar, who kindly shared his proposal entitled as "Dryfo Urbanism" for the purpose of this book. The result of the collaboration has been presented in World Design Summit in October of 2017 in Montreal, Canada; entitled as *Dryfo Urbanism Urbanization Through Forestation - Holistic Design Solutions* by Ishaan Kumar and Samaneh Nickayin.

[5]PROEXPORT - Ministerio de Comercio, Industria y Turismo, (2013) reported that during last six years, 1,906,909 international and close to 6,000,000 domestic tourisms have arrived in Colombia.

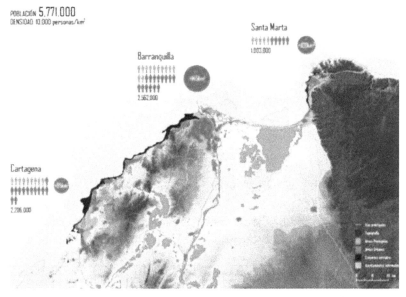

Figure 5.7 ©Diego Bermúdez, Chieh Huang (May 2015), Urbanization trends from Cartagena to Santa Marta. Landscape Department, University of Pennsylvania. Instructors: Richard Weller and David Gouverneur.

characterized by a dramatic shift from a rural/agricultural-based economy to one supported by industry, technology, and services.[6]

Due to wealth disparities, the population who migrated to these cities and towns self-constructed their urban habitats. With a steady population growth rate composed of 70% informal settlements, Cartagena epitomizes the global force of self-constructed cities lacking any semblance of natural amenities. Such urbanization has historically encroached on fragile and valuable ecosystems such as the dry forest and mangroves (Figure 5.8) and lacked any pretence to intentionally define the public realm and use landscape as an amenity to moderate harsh environmental and social conditions.

The ecological functions of tropical dry forest ecosystem and mangroves ecosystems are deeply connected. During the dry season, which has a length

---

[6]These migratory trends were also enhanced by high levels of violence, derived from historical social inequalities, exacerbated by the assassination of the Presidency in 1948: Jorge Eliécer Gaitán. This event led to the rise of guerrilla movements, and, over time, it embarked in the production and commercialization of drugs. The clandestine laboratories were located in remote areas and high levels of violence in such areas, forced massive displacements, and migration to the larger cities (Angrist and Kugler, 2008).

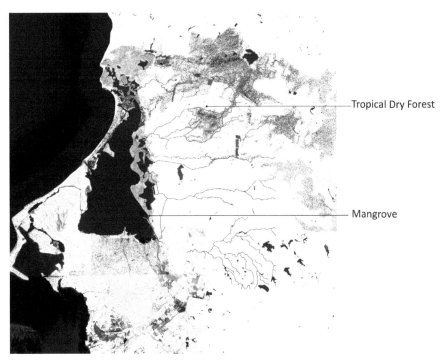

Figure 5.8 ©Boyang Li, Xinyi Ye, Tianjiao Zhang (May 2017), Dry forest ecosystem and mangroves ecosystems. Landscape Department, University of Pennsylvania. Instructors David Gouverneur.

of 7-8 months, tropical dry forest vegetation shed their leaves (Rzedowsky, 1978) to increase vegetation's hydric and mineral nutrients deficit (Bullock et al., 1995). The mangroves (characterized by sandy soils) are negatively affected by droughts, which cause higher soil salinity due to their low water and nutrient retention capacity (Jaramillo, 2014). Furthermore, significant precipitation could cause a diminution in mangrove extensions and alters their floristic composition (Bernal et al., 2016). Therefore, the seasonality that the tropical dry forest presents becomes a dominant ecological force to protect the seasonal streams, soils, and nutrients that feed the marshes where the mangroves grow (Galeano et al., 2017).

Along the coast, the growth of informal settlements has gradually infilled the marshes, eliminating the mangroves (Rangel-Buitrago et al., 2015), while in hinterlands, it encroached on the ecosystem of dry forest, mainly due to the fires and conversion to agriculture and urbanized area (Miles et al., 2006)

Figure 5.9   ©Boyang Li, Xinyi Ye, Tianjiao Zhang (May 2017), Urbanization trends of Cartagena. Landscape Department, University of Pennsylvania. Instructors David Gouverneur.

(Figure 5.9). Currently, 66% of dry forests have been converted into land uses through the Americas (Trejo and Dirzo, 2000).

This part represents a novel strategy of planning for the self-constructed city of Cartagena, which identifies land adequate for development and incorporates urban forestry as green infrastructure. "Dry Forest (Dryfo) Urbanism" proposes an orchestrated planting of the dry forest as the "Informal Armature" and departure point for future growth. Leveraging the existing vegetation of the grazing land, Dryfo Urbanism extends these lines throughout the expansion zone of Cartagena, incorporating a process of forestation to reserve future mobility corridors and provide severely needed public space (Figures 5.10 and 5.11).

This proposal imagines a future for Cartagena where the dry forest provides space for communal interaction in such a segregated society and increases tree canopy and biodiversity in a self-constructed context where citizens can enjoy their contextual ecosystem as a component of their

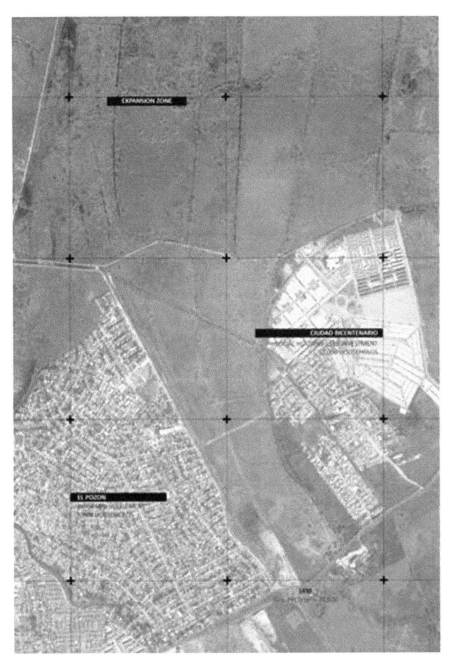

Figure 5.10 ©Ishaan Kumar. Cartagena contrasts. The existing community of informal settlement and social housing projects are in complete isolation.

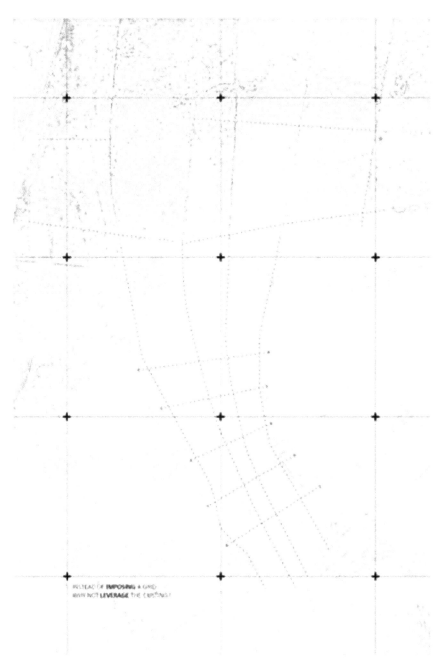

Figure 5.11   ©Ishaan Kumar. Dryfo Urbanism leverages the exiting vegetation of the grazing land to establish the urban framework and stretches the two communities together.

daily lives. David Gouverneur (2014) expresses that "throughout history, unplanned and informal processes have played an important role in shaping cities. It is when this informal mode of city making increases in scale that it causes health hazards, social segregation, social unrest, and ineffective governance". Dryfo Urbanism provides a framework for informal and formal infill, where the landscape is an integral part of the public realm, and establishes the "informal armature" for citizens to shape their own habitat and enjoy the benefits of shade and undeniable beauty.

The proposal emphasises the role of "landscape as protagonist", which seeks to recover the tree canopy of dry forest, attempts to restore the ecological loss of trees in the hot Caribbean environment, and integrates novel types of public space in informal contexts.

Colombia is the second most biodiverse country in the world (Rodríguez Becerra, 2001; Arbeláez- Cortés et al., 2015). Such biodiversity does not consist of just flora and fauna but encompass migratory people as the population of an internally displaced person (IDP)[7] (UNHCR, 2016). With adequate education, participation, and planning, such population who benefit from skills ranging from agriculture, forestry, and even artisanal craftwork could be considered as good "steward" (McHarg, 1992)[8] of landscape and dry forest.

Dryfo Urbanism proposes a selection of three families to strengthen and extend the existing lines of vegetation from the grazing territory. It establishes the urban framework for upcoming urbanization between defined corridors.

Such a proposal installs a botanical education among citizens and provides food, fruit, medicine, and wood to mitigate poverty among the inhabitants. Flanking each corridor could be a series of plazas acting as dense islands of each family (Figure 5.12).

With a diminishing degree of design and control approaching the arroyos and bodies of water, the evolution of these spaces could integrate government and local needs and skillsets. To reserve the land for these amenities, the government could work with local leaders to plant wood trees as a land bank system, chopping them down for future construction and furniture material as the city expands. Replacing these patches could be a network of community

---

[7]"Internally displaced people (IDPs) have not crossed a border to find safety. Unlike refugees, they are on the run at home, IDPs stay within their own country and remain under the protection of its government, even if that government is the reason for their displacement. As a result, these people are among the most vulnerable in the world" (UNHCR, 2016).

[8]See also Stewardship Now? Reflections on Landscape Architecture's Raison d'être in the 21st Century by Richard Weller, *Landscape Journal* 33(2), p. 1-24.

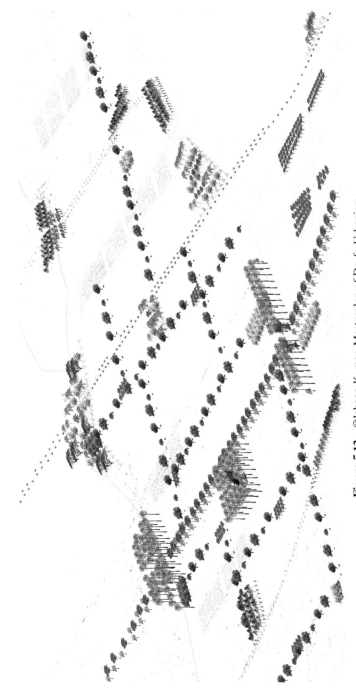

Figure 5.12   ©Ishaan Kumar, Masterplan of Dryfo Urbanism.

centres, schools, and agricultural fields generating urban nodes while serving as stewards over these invaluable assets of plazas and parks (Figures 5.13 and 5.14).

With the necessary North-South corridors established, the dry forest families could be threaded together by selecting shade and fruit trees, providing pedestrian corridors with invaluable shade (Figures 5.15 and 5.16). Intersecting pedestrian streets could have small pocket plazas highlighting a distinct colour of the Fabaceae family. Instead of honouring individuals and conquistadores such as Simon Bolivar, these plazas could be named after the tree itself, instilling a botanically educated city where local appropriation and sense of place is generated through the humble planting of trees. To achieve a higher degree of resolution on the character of the corridors, the Bombacaceae family is selected for the cultural association of the Macondo tree. Globally recognized as Colombia's most distinguished author, Gabriel Garcia Marquez named Macondo the location of his fictional city in the infamous 100 Years of Solitude. Unfortunately, without knowing that Macondo was a tree, the tree has become inevitably tied to solitude.

Dryfo Urbanism hopes to reverse this reality by featuring the Macondo in the Bombacaceae corridors. By integrating Macondo trees into the daily lives of commuters, the 50-meter tree becomes emblematic of flows, movement, and, most importantly, unwavering unity. Raised platforms could even elevate the Macondo to be placed on a symbolic and poetic pedestal and feature its elephant-like base at eye level. Pairing it with the undulating base of the Ceiba, the design of the plazas and parks could be seamlessly integrated, responding to the varying characteristics of the individual trees.

Dryfo Urbanism is not the traditional urban design of imposition over a territory but a "Landscape narrative" (Potteiger and Purinton, 1998)[9] that highlights what is already there, expressing inherent social, cultural, ecological, historical, and political voices of Cartagena. By guaranteeing mobility corridors and public space, Dryfo Urbanism blends forestation and urbanization, where "Landscape [acts] as Urbanism" (Waldheim, 2016) and transform the informality of self-constructed city to respect the ecosystem of dry forest, restoring and balancing the fragile ecosystem and go beyond the short-term, market-driven needs, looking from a broader-scale to response to the emergency and disaster of deforestation caused by unplanned informality.

---

[9]"Narratives [. . .] intersect with sites, accumulate as layers of history, organize sequences, and inhere in the very materials and processes of the landscape. In various ways, stories 'take place'". Potteiger, M., Purinton, J. (1998). *Landscape Narratives: Design Practices for Telling Stories.* New Jersey: John Wiley & Sons. p. 5.

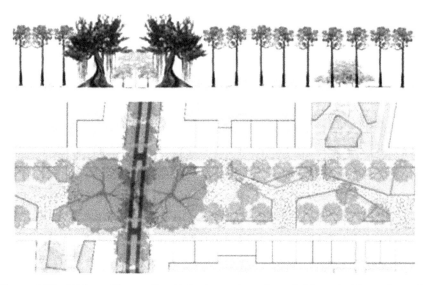

Figure 5.13 ©Ishaan Kumar. Detailed plan and section of Macondo Plaza along the Bombacaceae corridor with intersecting Fabaceae plazas and surrounding commercial node.

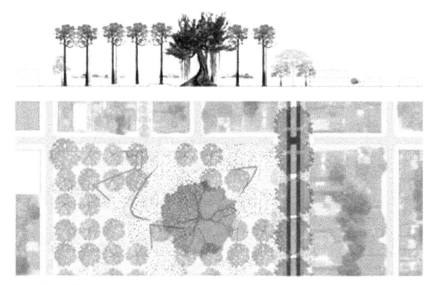

Figure 5.14 ©Ishaan Kumar. Detailed plan and section of the larger plaza along the Bombacaceae corridor providing urban relief integrating the intricacies of the individual trees.

Figure 5.15 ©Ishaan Kumar - Public space around Bombacaceae corridor.

Figure 5.16 ©Ishaan Kumar - Public space and city life

Eco-poverty[10] in mangroves: San José de Chamanga[11]

The main concern in self-constructed cities is poverty. Even though there is no unique way to define poverty, it is commonly defined as deprivation in well-being in monetary, powerlessness, social exclusion, vulnerability, and lack of opportunity and resources (Ruggeri Laderchi et al., 2003).

According to Brundtland Report, "[...] Poor people are forced to overuse environmental resources to survive from day to day, and the impoverishment of their environment further impoverishes them, making their survival even more difficult and uncertain" (WCED, 1987).[12] Since the rise of idea, sustainable development, environmental degradation, and poverty have been considered reciprocating (Coward et al., 1999; Scherr, 2000; Gray and Moseley, 2005; DeClerck et al., 2006).

The goals encompassed in this essay relate to mangroves' contribution to reducing poverty among the poor residents of the self-constructed city of San José de Chamanga in Ecuador.

Mangroves provide essential areas of hatching for brackish and open water fish and contribute to poor dwellers of coastal areas that often rely on fishing and other economies based on the natural environment. The destruction of the fragile ecosystem of mangroves (due to urbanization, industrialization, and deforestation) has further impoverished the community of Chamanga (Figures 5.17 and 5.18).

Mangroves are among the most valuable and productive ecosystems of coastlines (Bao et al., 2013). For instance, due to their demands for nutrients, they play essential roles in removing pollutants, heavy metals, and pesticides (Kristensen et al., 2008; Reef et al., 2010; Adame and Lovelock, 2011; Bayen, 2012) and are considered as tolerant plants for wastewater effluent (Yang et al., 2008; Huang et al., 2012). Furthermore, mangroves deter the advancement of seawater onto sweet water bodies, preserving biomes, diminishing the impact of sea-level rise, storms, tsunamis, and erosion (Tanner et al., 1991; Alongi, 2008; Zhang et al., 2012; Aung et al., 2013). The devastation of

---

[10]See also Nofroni, L. (2017). *Paesaggi delle eco-povertà nel mediterraneo. Il paesaggio come strumento di osservazione e di proiezione strategica per il superamneto delle inquità eco-sociali.* PhD thesis. Sapienza, Università di Roma.

[11]It is based on academic proposal of Aubrey Jahelka and Shuwen Ye, entitled as "San Jose de Chamanga, Rebuilding for Resilient Landscapes", conducted in the Landscape Architecture Department at the University of Pennsylvania in 2017, who kindly shared their proposal for the purpose of this research.

[12]Report of Brundtland well-known also as our common future, p. 27.

Figure 5.17 ©Aubrey Jahelka—San José de Chamanga.

Figure 5.18 ©Shuwen Ye—San José de Chamanga.

mangroves has exacerbated the landslide and erosion along the coastline of Chamanga.

San José de Chamanga experiences a 2.5-meter fluctuation in tides. Moreover, natural disasters (earthquake, flooding, and landslide) hit frequently this area. The government suggests that the community relocated further upland to remove them from risk. Nevertheless, 80% of the residents rely on access to the water for their livelihood, not to mention the storage of boats, nets, and motors. This new location completely disregards the development and heart of Chamanga, which is a fishing community along the waterfront. Without the prevalence of a fishing-based community, it is unclear what economic opportunities exist for the residents. Furthermore, there is no reasonable enforcement mechanism to prevent people from rebuilding the original informal fabric along the coastline.

If the mangroves were to be restored, the trees themselves would act as a natural amplitude dampener to tsunami and sea-level rise; simultaneously, it preserves economic opportunities for the residents.

Many countries pursue various forms of silvo-aquaculture as a win-win strategy for conserving mangroves and fostering economy and employment among low-income dwellers of the coastal community (Sukardjo, 1989; Primavera, 2000; Oswin and Ali-Hussain, 2001; Suh, 2012).

The proposal of Jahelka and Ye is based on the Silvofishery technique, which integrates mangroves with brackish aquaculture. They proposed the Chinampa technique as the restoration strategy for mangroves. The system of chinampas is a traditional Mexican agrarian technique, which includes a series of the agronomic, fishery, and forestry activities. The chinampas are raised bed soils constructed by sediments of excavated place, creating a system of islands separated by channels (Merlín-Uribe et al., 2012) (Figure 5.19).

Furthermore, such a proposal creates a commercial corridor along the waterfront to conserve the waterfront as the spine of the town and the centre of social life, without a drastic alternation in the life of inhabitants (Figure 5.20). This proposal is a holistic vision that intertwines the restoration of mangroves, economic and cultural aspects, and informality (Figure 5.21).

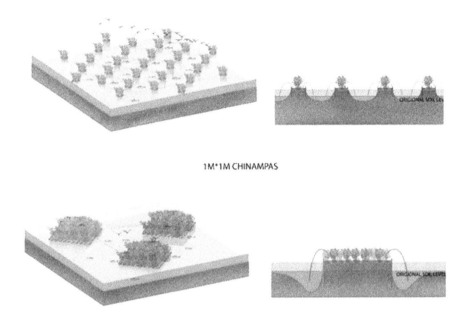

1M*1M CHINAMPAS

10M*10M CHINAMPAS

Figure 5.19 Islands of Chinampas. ©Aubrey Jahelka, Shuwen Ye.

It is an efficient example of forestry that Vandana Shiva (1993) calls "Life-enhancing" forestry. In her book *Monocultures of the Mind, Perspectives on Biodiversity and Biotechnology*, she specifies that:

> "[...] The survival of [...] forests depends on the survival of human societies modelled on the principals of the forest [...], hidden in the lives and beliefs of the forest peoples of the world [...] The life-enhancing [forestry] emerges from the forest and forest communities [and] creates a sustainable, renewable forest system, supporting and renewing food and water systems. The maintenance of conditions for renewability is the primary management objective [...]".

Figure 5.20   Masterplan. ©Aubrey Jahelka, Shuwen Ye.

Figure 5.21 ©Aubrey Jahelka, Shuwen Ye.

References

Adame, M.F., Lovelock, C.E. (2011). Carbon and nutrient exchange of mangrove forests with the coastal ocean. *Springer: Hydrobiologia*, 663(1), p. 23–50.

Alongi, D.M. (2008) Mangrove forests: resilience, protection from tsunamis, and responses to global climate change. *Estuar Coast Shelf Sci* 76(1), p. 1–13.

Arbeláez- Cortés E., Torres, M.F., López Álvarez, D., Palacio Mejía, J.D., Mendoza, A.M., Medina, C.A. (2015). Colombian frozen biodiversity - 16 years of the tissue collection of the Humboldt Institute - La biodiversidad congelada de Colombia: 16 años de la colección de tejidos del Instituto Humboldt. *Acta Biológica Colombiana,* 20(2), p. 163-173.

Aung, T.T., Mochida, Y., Than, M.M. (2013) Prediction of recovery pathways of cyclone-disturbed mangroves in the megadelta of Myanmar. *For Ecol Manage* 293, p. 103–113.

Bao, H., Wu, Y., Unger, D., Du, J., Herbeck, L.S., Zhang, J. (2013). Impact of the conversion of mangroves into aquaculture ponds on the sedimentary organic matter composition in a tidal flat estuary (Hainan Island, China). Elsevier: *Continental Shelf Research.* 57, p. 82–91.

Barreda-Bautista, B., López-Caloca, A.A., Couturier, S., Silván-Cárdenas, J. L. (2011). Tropical Dry Forests in the Global Picture: The Challenge of Remote Sensing-Based Change Detection in Tropical Dry Environments. *Planet Earth 2011_Global Warming Challenges and Opportunities for Policy and Practice.* p. 232-256.

Bayen, S. (2012). Occurrence, bioavailability and toxic effects of trace metals and organic contaminants in mangrove ecosystems: a review. *Elsevier: Environment International*, 48, p. 84–101.

Bernal, G., Osorio, A.F., Urreg, L., Peláez, D., Molina, E., Zea, S., Montoya, R.D., Villegas, N. (2016) Occurrence of extreme oceanic events and their impact on the ecosystems. *Elsevier: Journal of Marine Systems*, 164, p. 85–100.

Bullock, S. Mooney, H. and Medina, E. (1995). *Seasonally dry tropical forests.* Cambridge: Cambridge University Press. p. 450.

Coward, E.W., Oliver, M.L., Conroy, M.E. (1999). Building natural assets: re-thinking the Centers, Natural Resources Agenda and its links to poverty alleviation. In: *Conference on Assessing the Impact of Agricultural Research on Poverty Alleviation*. September 14–16, San Jose, Costa Rica.

DANE-National Department of Statistics (2013). *General Census of Colombia y* (Report 1). [online]. Bogota: DANE, Santa Fe de Bogota. Available at: https://www.dane.gov.co/files/investigaciones/pib/depar tamentales/ingles/Methodology%20of%20Regional%20Accounts.pdf [Accessed 1 July 2017].

DeClerck, F., Ingram, J.C., Rumbaitis del Rio, C.M. (2006). The role of ecological theory and practice in poverty alleviation and environmental conservation. *Front Ecol Environ*, 4(10), p. 533–540.

FAO (2003). *State of the World's Forests 2003. Food and Agriculture Organization of the United Nations*. [online]. Available at: http://www. fao.org/docrep/005/Y7581E/Y7581E00.HTM [Accessed 3 August. 2017].

FAO (1993), *Management and conservation of closed forests in tropical America*. [online]. Rome: FAO Forestry Paper No.101. Available at: http://www.fao.org/docrep/016/ap420e/ap420e00.pdf [Accessed 25 Apr. 2017].

Galeano, A., Urrego, L.E., BoteroV., Bernal, G. (2017). Mangrove resilience to climate extreme events in a Colombian Caribbean Island. *Springer: Wetlands Ecol Manage.*

Gouverneur, D. (2014). *Planning and design for future informal settlements*. New York: Routledge.

Grahn, L.R. (1996). Cartagena. In: *Encyclopedia of Latin American History and Culture*, 1. New York: Charles Scribner's Sons. p. 581.

Gray, L.C, Moseley, W.G. (2005). A geographical perspective on poverty–environment interactions. *The Geographical Journal*, 171(1), p. 9–23.

Huang, Q., Liu, Y., Zheng, W., Chen, G. (2012). Phytoplankton community and the purification effect of mangrove in the mangrove plantation-aquaculture coupling systems in the Pearl River Estuary. *Elsevier:* Procedia Environmental Sciences, 15, p. 12–21.

IBDF. (1981). *Plano de Manejo Parque Nacional da Tijuca*. Ministério do Meio Ambiente, FBCN. p. 39–45.

Jaramillo, D.F. (2014). El Medio Físico Edáfico. In: *El Suelo: Origen, propiedades, Espacialidad*. 2nd ed. Medellín: Universidad Nacional de Colombia, p 137.

Kristensen, E., Bouillon, S., Dittmar, T., Marchand, C. (2008). Organic carbon dynamics in mangrove ecosystems: a review. *Elsevier: Aquatic. Botany.* 89, p. 201–219.

McHarg, I.L. (1992). *Design with Nature*. New York: John Wiley and Sons.

Merlín-Uribe, Y. , E. González-Esquivel, C. , Contreras-Hernández, A., Zambrano, L. , Moreno-Casasola, P., Astier,M. (2012). Environmental and socio-economic sustainability of chinampas (raised beds) in Xochimilco,

Mexico City. *International Journal of Agricultural Sustainability* ,11(3), p. 216-233.

Miles, L., Newton, A.C., DeFries, R. S., Ravilious, C., May, I., Blyth, S., Kapos, V. & Gordon, J. (2006). A global overview of the conservation status of tropical dry forests. *Journal of Biogeography*, 33(3), p. 491-505.

Moya Pons, F. (2007). *History of the Caribbean: Plantations, Trade, and War in the Atlantic World.* New Jersey: Markus Wiener Publisher.

Oberndorf, R., Durst, P., Mahanty, S., Burslem, K., Suzuki, R. ed. (2007). *A Cut for the Poor- Proceedings of the International Conference on Managing Forests for Poverty Reduction: Capturing Opportunities in Forest Harvesting and Wood Processing for the Benefit of the Poor.* Report No. 19. Ho Chi Minh City, Viet Nam 3-6 October 2006. Bangkok: FAO and RECOFTC, Report No. 19. Available at: http://www.fao.org/docrep/0 10/ag131e/ag131e00.htm [Accessed 26 June 2017]

Oswin, S.D., Ali-Hussain, S. (2001). Integrated mangrove Shrimp Silvofisheries: a pioneer organic shrimp culture model in India. *Aquaculture Asia*, 4(4), p. 5–9.

Parker, R., Kreimer, A., Munasinghe, M. (eds.) (1995); *The International Decade for Natural Disaster Reduction(IDNDR). Informal settlements, environmental degradation, and disaster vulnerability: The Turkey case study.* [online] Washington, D.C.: The World Bank. Available at: http: //documents.worldbank.org/curated/en/693531468761060935/Inform al-settlements-environmental-degradation-and-disaster-vulnerability-the-Turkey-case-study [Accessed 31 July 2017].

Potteiger, M., and Purinton, J. (1998). *Landscape Narratives: Design Practices for Telling Stories.* New Jersey: John Wiley & Sons.

Primavera, J.H. (2000) Integrated mangrove-aquaculture systems in Asia. *Integrated Coastal Zone Management.* p. 121–130.

PROEXPORT e Promoción de Turismo, Inversión y Exportaciones (2013). *Informe derendición de cuentas para el sector comercio, industria y turismo.* Bogota: Ministerio de Comercio, Industria y Turismo.

Rangel-Buitrago, N., Anfuso, G., Williams, A.T. (2015). Coastal erosion along the Caribbean coast of Colombia: Magnitudes, causes and management. *Elsevier: Ocean & Coastal Management*, 114, p. 129-144.

Rangel-Buitrago, N., Correa, I., Anfuso, G., Ergin, A., Williams, A.T. (2013). Assessing and managing scenery in the Caribbean coast of Colombia. *Elsevier: Tourism and. Management*, 35, p. 41-58.

Reef, R., Feller, I.C., Lovelock, C.E. (2010). Nutrition of mangroves. *Tree Physiology.* 30(9), p. 1148–1160.

Rodríguez Becerra, M. (2001). La biodiversidad en Colombia. [online]. Available at: http://www.manuelrodriguezbecerra.org/bajar/politicae.pdf [Accessed 20 July 2017].

Ruggeri Laderchi, C., Saith, R., Stewart, F. (2003). Does it Matter that we do not Agree on the Definition of Poverty? A Comparison of Four Approaches. *Oxford Development Studies,* 31(3), p. 243- 274.

Rzedowsky, J. (1978). *Vegetación de México.* Mexico: Editorial Limusa S. A.

Scherr, S. J. (2000). A downward spiral? Research evidence on the relationship between poverty and natural resource degradation. *Elsevier: Food Policy,* 25, p. 479–498.

Shiva, V. (1993). Monocultures of the Mind, Perspectives on Biodiversity and Biotechnology. London: Zed Books, p.19-20.

Suh, J. (2012). The Past and Future of Community-Based Forest Management in the Philippines. *Philippine Studies: Historical and Ethnographic Viewpoints. Ateneo de Manila University.* 60(4), p. 489-511.

Sukardjo, S. (1989) Tumpang sari pond as a multiple use concept to save the mangrove forest in Java. *Biotrop Special Publication,* 37, p.115–128.

Tanner, E.V.J., Kapos,V., Healey, J.R. (1991) Hurricane effects on forest ecosystems in the Caribbean. *Biotropica* 23(4), p.513–521.

Trejo, I., Dirzo, R. (2000). Deforestation of seasonally dry tropical forest a national and local analysis in Mexico. *Biological Conservation,* 94(2), p.133-142.

UN-Habitat (2003 a). *Slums of The World: The Face of Urban Poverty in The New Millennium?.* [online]. Nairobi: UN-HABITAT, p.30. Available at: https://archive.org/details/SlumsOfTheWorldTheFaceOfUrbanPoverty InTheNewMillennium [Accessed 10 June 2017]

UN-Habitat (2003 b). *The Challenge of Slums, Global Report on Human Settlements.* [online]. Revised and updated version (April 2010), London: Earthscan Publications Ltd. Available at: https://unhabitat.org/wp-content /uploads/2003/07/GRHS_2003_Chapter_01_Revised_2010.pdf [Accessed 30 July 2017].

UN-Habitat (2013). *The State of the World Cities Report 2012/13.* [online]. New York: Routledge. Available at: https://mirror.unhabitat.org/pmss/ge tElectronicVersion.aspx?nr=3387&alt=1 [Accessed 15 July 2017].

UNHCR -United Nations High Commissioner for Refugees (2016). *Colombia Situation update.* [online]. Available at: http://www.refworld .org/docid/58627cd24.html [Accessed 3 August 2017]

Waldheim, C. (2016). Landscape as Urbanism: A General Theory. New Jersey: Princeton university press.

WCED_ World Commission on Environment and Development (1987). Our Common Future. *Oxford: Oxford University Press.* p. 27.

White, A., Martin, A. (2002). *Who Owns the World's Forests? Forest Tenure and Public Forests In Transition.* Washington, DC.: Center for International Environmental Law.

World Bank (2008). *Urban Poverty, World Bank Urban Papers, 2008.* New York: Oxford University Press.

Yang, Q., Tam, N.F.Y., Wong, Y.S., Luan, T.G., Su, W.S., Lan, C.Y., Shin, P.K.S., Cheung, S.G. (2008). Potential use of mangroves as constructed wetland form municipal sewage treatment in Futian, Shenzhen, China. *Elsevier: Marine Pollution Bulletin.* 57(6-12), p.735–743.

Zárate, L. (2016). They are Not "Informal Settlements"—They are Habitats Made by People. *The nature of cities* [online]. Available at: https://www.th enatureofcities.com/2016/04/26/they-are-not-informal-settlements-they-are-habitats-made-by-people/ [Accessed 6 June. 2017]

Zhang, K., Liu, H., Li, Y., Xu, H., Shen, J., Rhome, J., Smith, T.J. (2012) The role of mangroves in attenuating storm surges. *Estuar Coast Shelf Sci* 102–103, pp.11–23.

6

Towards a Planetary Vision: Conclusion

Towards a planetary vision

"Action has meaning only in relationship, and without understanding relationship, action on any level will only breed conflict. The understanding of relationship is infinitely more important than the search for any plan of action".

Jiddu Krishnamurti

This book set out to identify frameworks, methods, and tools that can help decision-makers and experts to be more effective in making and managing forests in/near *civis* habitats. The book has sought to define a new paradigm of forestation as a driver of urban development and not a balance to its adverse effects, basing on holistic approaches.

The findings affirm that the efficiency of urban forestry can be guaranteed if and only if the traditional limits between urban, peri-urban, and rural forest shift to a holistic framework. Digging into different theories reveals that "Limit exists only in mind"[1]! If we mark a circle around an ant, it goes into a loop inside the circle because it perceives the line as a limit, albeit it does not exist physically.

Fortunately, we are not an ant, we are human, and our collective habits of thoughts and actions helped us to overcome the limits and boundaries, as Hunt (2012) affirms the margins in landscape studies have vanished.[2] In such embodied system, designing cities means even inventing agriculture (Zagari, 2013), as everything "is hitched to everything else in the Universe"[3] (Muir, 1911).

---

[1] Alan Cohen writes "All limits exist only in the mind, and it is only in the mind that they can be overcome".

[2] Hunt, D.J. (2012). *Sette lezioni sul paesaggio.* Trans. Morabito, V. Melfi: Libria, p. 19.

[3] Cited by Giffor, T. (2006). *Reconnecting with John Muir: Essays in Post-Pastoral Practice.* Georgia: University of Georgia, p. 25.

Humanity went beyond the duality between city and nature, and as Richard Forman says, "you can have a small positive impact in a city centre, but if you want to have a big impact, go out into this dynamic urban edge where solutions really matter for both nature and people. Antoni Gaudi designed the magical Parque Güell out on the edge of Barcelona, and Frederick Olmsted's Emerald Necklace was created on the dynamic edge of Boston".[4]

In the era of globalization, when the "urban" edge does not exist anymore, this book sought to point out, through different holistic theories, the reasons why it is unavoidable to shift the paradigm of "urban" forest(ry) towards a planetary vision.

6.1 "If God is Now 'Green'"
Theoretical recommendation for today

In the era of Anthropocene, with its endless "ecological hell", Richard Weller (2015) writes that:

> "If God is now "green" and Adam were to be evicted from Paradise again, then his punishment would not be to convert wilderness into farms, but farms back into wilderness. Eve, renowned for her scientific curiosity, would lead the way".

In the face of the environmental apocalypse, greening became a "universal moral imperative" (Wilson, 2006) of many scientists, governments, and citizens; even Pope Francis has written the encyclical letter *Laudato Si* on the care of our common home. All around the world, our necrotic habitats' physical and social quality have turned into the preeminent call. The wide range of greening concepts and set of practices have been produced, reproduced, and transformed to the point that confusions have arisen. The only real fact is that tree plantation is our "divine punishment" in the "Sixth Extinction"[5] era.

The affirmation that the entire planet is urbanized implies new paradigm of the "edge" of the urban area. Therefore, as the first paradigm, the edge of urban or peri-urban forest(ry) should shift to a planetary threshold.

---

[4]See "In conversation with Richard T.T. Froman" in *LA+ Interdisciplinary Journal of Landscape Architecture* 1. p. 114-116.

[5]See Elisabeth Kolbert. (2014). *The Sixth Extinction: An Unnatural History.* New York: Henry Holt & Company.

Figure 6.1 ©NASA Earth Observatory images by Joshua Stevens, using Suomi NPP VIIRS data from Miguel Román, NASA's Goddard Space Flight Center.

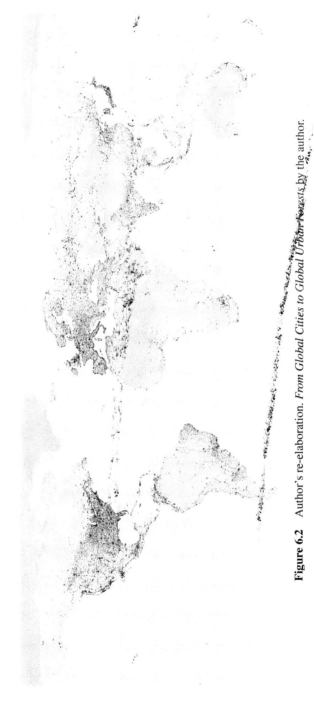

Figure 6.2   Author's re-elaboration. *From Global Cities to Global Urban Forests* by the author.

The proposal of "planetary" forest(ry) (Figures 6.1 and 6.2) not only tends to *re*imagine new physical and geographical edge of urban forest(ry), but it also attempts to broaden the visions towards the role that is assumed to shape the context in which it takes part. The holistic theories and approaches have broadened the horizons of our knowledge. Richard Forman and Wu (2016), by their holistic suggestion on a global scale, have already defined the "habitable zone" "where to put next billion people"; likewise, it means "where to put next billion trees" for the emerging urban areas. The global trends of urbanization are predictable; thus, we should interweave the trends of forestation. Thinking holistically and acting locally, and as Richard Tarnas (2006) remarks:

> "[…] we must go not only high and far but down and deep. Our world view and cosmology, which defines the context for everything else, is profoundly affected by the degree to which all our faculties – intellectual, imaginative, aesthetic, moral, emotional, somatic, spiritual, relational – enter the process of knowing. How we approach "the other," and how we approach each other, will shape everything, including our own evolving self and the cosmos in which we participate[6]".

Today, the scale of cities has changed; satellite images reveal a continuous chain of metropolitan areas connected by transport links, and the management of such patchworks should be done holistically. Holistic thinking reveals the best arrangement of land systems for human and nature.

As discussed, the trends of urbanization are detectable; in line with such trends, we can define the role that forests can assume. For instance, mangroves or dry forests are not typically considered urban forests, but the way urban areas and people influence them incites a new understanding of their role in urban areas. In his holistic proposal, Ishaan Kumar intertwined the expansion trends of informal settlements with the conservation of dry forests, where the dry forests built the "informal armature" for upcoming urbanization.

Today, innovative approaches that embrace the dynamic transformation of urban areas (like expansion or shrinkage) and nature conservation are sorely needed. Fuelled by a burgeoning population and economic tendency, the process of urbanization shapes the territories, creating enmeshed systems

---

[6]Tarnas R. (2006). *Cosmos and Psyche: Intimations of a New World View.* New York: Viking.

of cities and nature. In such scenarios, the big challenge is the interaction of these two systems, which creates a "whole" one system, where physical boundaries do not make sense.

Designing with the "process" (and not "boundaries") of urbanization "invites improvisation and chance, [...] opening up a host of other completely unexpected relationships and outcomes" (Corner, 2016).

6.2 Rethinking the Urban Forests for the 21st Century
Practices

Albert Einstein writes: "we cannot solve our problems with the same thinking we used when we create them". Architect and planner Dilip da Cunha[7] explains similar concepts, talking about flooding; he argues that the flooding is a definition of human, and when we design a line and the river overcomes that line, it is considered "flooding".

Similarly, in the new era of urbanization, the concept of urban forest(ry) cannot follow the previous arbitrary definition and function. Today, the role of urban forestry is mainly debated for air pollution, but its effectiveness in the future is debatable. The next billion people need food, water, and shelter. It means that more soil will be consumed only to provide the principal amenities, more ecological footprint will produce, and "as this century unfolds there will not be enough land to utilize forestry as the single mechanism for carbon sequestration"; according to this statement, the Earth's biological capacity of next billion people is calculated as 670% of the Earth's current capacity[8] (Weller et al., 2017).

The growth of population and climate change is the main threats in the next century. Global warming will lead to both desertification and flooding in the north. In such a scenario, the role of urban forestry in the next century might not be merely Carbone sequestration or compensation to ecological footprint. It might create a "habitable zone" where the new settlements could erect, and humanity could survive.

---

[7]See Mathur, A., da Cunha, D. (2001). *Mississippi Floods: Designing a Shifting Landscape.* New Haven: Yale University Press.

[8]The research *Atlas for the End of the World* conducted at the Landscape Department of University of Pennsylvania, by Richard Weller (2017) and his group by visualizations design, illustrate the ecological footprint of 10 billion people and corresponding required forest in order to sequester the carbon emission. The details of this research are available at:http://atlas-for-the-end-of-the-world.com/datascapes/biocapacity.html

To be clear, let us refer to the example of the *Great Green Wall* of Africa. This colossal project of FAO is not classified as an urban forestry project, albeit it is expected to benefit rural and agricultural development. If so, within 40 years, when the forest is grown, probably an additional "great wall of settlements" would rise on the edge of the green wall for dwellers who contribute to and benefit from the services provided by the green wall. Then, why should it not be conceived and designed from the perspective of an "urban" forest right now?

Similarly, various large-scale landscape connectivity projects seek to mitigate species' habitats, reconnecting fragmented ecosystems and ensuring the migration of species all around the globe. With the upcoming challenges in the next century, we might rethink these unprecedent scales of projects as the corridors for species and as liveable areas for human species, reflecting on the possibility of new settlements on their borderline.

Many human-planted forests have been realized in treeless areas during history, with different scales and purposes. One case in point is the Nebraska National Forest, which was an experimental project. Today, the effort of landscape architect Horace Cleveland appears as the largest human-planted forest in the United States. Other similar prototypes with smaller scales are the historical Persian gardens of Shazdeh Mahan in Kerman (Iran) and the Lost Paradise in Neyshabur, which, thanks to the *qanat* system, create suitable microclimates for the trees in a dry context of the desert.

Even though these examples are not classified as forestation in the urban area, their replication in other areas could create a new habitable zone for future urban settlements as a win-win strategy for nature and human.

In short, the concept of *préverdissement* of Guinaudeau (1987) might be newly adopted on a global scale to convert the unliveable land to liveable land for the next billion people. It is necessary to construct many Parc Départemental du Sausset, and many landscape architects like Corajoud are required to work "as Urbanist of Our Age" as "responsible for the integration of civil infrastructure, [...] and environmental improvement" (Waldheim, 2016). Furthermore, if the human being changes his sedentary habits to become nomadic one more time, as Maurizio Corrado (2012) asserts, then the path of architecture leads into the forests. Such a statement could open new debates on the role of urban forestry, not as a method to compensate for the human's ecological footprint but to create unexplored territories to inhabit.

Furthermore, desertification and flooding are two main challenges of unprecedented climate change in our habitats. Forestation is the resilience approach that can be adopted in such contexts. "When it comes to climate, and [its safeguard] timing is everything" (Figueres et al., 2017). The sense of time in afforestation is indispensable. It takes years to make a forest; planting today will provide benefits within years. To avoid the catastrophic influence of climate change and population growth, we should act in advance to be able to be effective.

Moreover, the scale of food crisis opens debates on "Sustainable Food Planning",[9] which will be one of the most compelling challenges for the upcoming century. Recently, it has been a call for proposals regarding Urban Food forests as a potential contributor to liveability in urban areas.[10] Due to the multi-functional character of the food crisis, experts from a wide range of backgrounds define the food policy. For instance, in Italy, thanks to the project of *Città del Vino*, the food policy assumes a new position in urban planning and land management (Marino, 2016).

Nonetheless, in many regions of rapid population growth, few lands are available for agriculture. Statistics represent that there is a shortfall in arable land for the subsequent populations at a global scale.[11] The example of Nemo's garden by Ocean Reef Group, which promotes food cultivation under the ocean, affirms the new challenges that humanity is facing to find the alternative solution to grow food in a responsible, small-footprint-on-earth kind of way.[12]

Determining an accurate distribution between agricultural land and forest is a central dilemma for the upcoming century. To illustrate, the advancement of forests in abandoned farmlands of mountains is one of the most clamorous arguments among experts; such phenomenon declined the arable land and agricultural production and influenced biodiversity and landscape (MacDonald et al., 2000). Along with these challenges, an increment trend towards sustainable forest products could reduce the pressures on the intensification of crop production. Thus, forestation in urban areas can play a crucial role in overcoming the crisis of food. The proposal

---

[9]"Sustainable Food Planning" has been launched for the 8th Annual Conference of the AESOP, 2017.See http://www.aesop-planning.eu/blogs/en_GB/sustainable-food-planning

[10]See https://www.journals.elsevier.com/urban-forestry-and-urban-greening/call-for-papers/special-issue-on-urban-food-forests-a-potential-contributor

[11]For details see: http://www.futuredirections.org.au/publication/the-future-prospects-for-global-arable-land/ https://data.worldbank.org/indicator/AG.LND.ARBL.HA.PC

[12]See http://www.nemosgarden.com/

of Chef Sean Sherman, who reclaimed the indigenous culinary culture of North America,[13] opens a new panorama in food production. Such suggestion reduces the necessity of farmlands and aims at new food plan, with attention to the world around as sources of ingredients and re-educating the people's palate.

Food production is one of the functions attributed to the forests in future urban areas; the discussed examples of mangroves in San José de Chamanga and the proposal of "luxury gradient" by FAO can optimize the performance of urban forests in line with food shortcoming.

To conclude, the triptych of urbanization, forestation, and food production should be planned and perceived in a holistic correlation to tackle land shortage in the 21st century.

6.3 "A Good Anthropocene"
Prospects

> "Humanity may destroy the possibilities for life on earth unless the freedom and power that we have acquired are channelled in new creative directions by a spiritual awareness and moral commitment that transcend nationalism, racism, sexism, religious sectarianism, anthropocentrism, and the dualism between human culture and nature. This is the great issue for the 1990s and the twenty-first century".

> Steven Clark Rockefeller

"Urbanization. Growth of population. Ecological hell. Food scarcity. Climate change. The age of Anthropocene." for clarity.rbanization. Growth of population. Ecological hell. Food scarcity. Climate change. The age of Anthropocene.

In the age in which human hands remake the Earth, Vandana Shiva states, "today, the challenge we face is the challenge of defending the future. In the past, land was colonized, people were colonized, but today, by the recklessness of greed the future itself is being colonized [...] with climate change[14]". The fear of the future and the global crisis has led to the

---

[13]Sean Sherman. (2017). *The Sioux Chef's Indigenous Kitchen*. Minneapolis: University of Minnesota Press.For further information, see http://sioux-chef.com/

[14]Interview with Vandana Shiva. Watch I *Nove Semi - L'india di Vandana Shiva* by Maurizio Zaccaro. Available at: https://www.youtube.com/watch?v=5oz8q1nfvSY

organization of projects like the *Svalbard Global Seeds vault*; a seed bank in the Arctic to ensure the preservation of seeds against global catastrophic risks, creating "[...] a hibernated Garden of Eden. A place where life can be kept forever, whatever happens in the world[15]".

Even though, in comparison with our ancestors, we have achieved the technologies and knowledge that permit us to support higher living standards with much lower per-capita impacts on the environment, the application of such knowledge does not seem to meet our expectations.

Nonetheless, a ray of hope still exists for our planet. It is foreseen that population size will reach the highest point in this century and then it will start to decline. Such a statement ensures new opportunities for a better future. The sprawl might stop, and the population will be stabilized. Hence, hopeful scenarios would open in management of lands.

A group of 18 environmental scientists, activists, scholars, and economists co-authored *An Ecomodernist Manifesto* (Asafu-Adjaye et al., 2015), affirming an optimistic vision for a future in which "humankind's extraordinary powers in the service of creating a good Anthropocene [...] "de-couple" human development from environmental impacts [...] allowing more room for non-human species [and] for nature".

If we overcome the actual crisis by the wisdom of a good Anthropocene, we can save the Earth and generate hopes for humanity's future, and all species on the earth investing in long-time scale.

Our world needs good Anthropocenes with holistic approaches that act for the seventh generation's benefits into the future.[16]

Currently, the "elements of a Good Anthropocene" exist in our world. The project of *Seeds of Good Anthropocenes* (Bennett et al., 2016) collects such prototype initiatives into a database to expand the range of ideas about environmental futures; among them, we can mention Food Forest Ketelbroek, Vertical Forest by Boeri architecture firm in Milan, Tree of 40 fruits by New York-based artist Sam Van Aken.[17]

McHarg placed great hope in the power of human as the "good Steward", who "green the earth, restore the earth [and] heal the earth". Landscape

---

[15]The statement of EU President José Manuel Barroso, during the opening ceremony of Svalbard Bank. See https://www.vg.no/nyheter/innenriks/en-frossen-edens-hage/a/518299/

[16]Refers to the constitution of the Iroquois: "Look and listen for the welfare of the whole people and have always in view not only the present but also the coming generations, even those whose faces are yet beneath the surface of the ground – the unborn of the future Nation".

[17]The projects of good Anthropocene are available at: https://goodanthropocenes.net/map-of-seeds/

designers, the "incurable optimistic species" as Franco Zagari defines, can mitigate such ecological hell. Regarding such capacity, Richard Weller (2014), writes *Reflections on Landscape Architecture's Raison d'être in the 21st Century*, underlining the importance of landscape urbanism at a global scale:

> "Landscape urbanism's most powerful insight is that 'the city' is ecological and that it is not a discrete object but now a global system without edge. The greater potential for landscape urbanism, therefore, is to scale up to Brenner's 'planetary urbanization' and link the McHargian tradition of large scale landscape planning with the global conservation and scientific community to help develop spatial plans that show how the ubiquitous forces of urbanization and its related industrial and agricultural infrastructure can be reconciled with biodiversity".

Landscape architects are good Anthropocenes, who benefit from a multi-scalar orientation with the ability to combine bottom-up and top-down approaches[18]; such abilities are crucial to address the forestation in line with other issues, like urbanization and food production.

Good Anthropocenes can answer such antagonistic issues because they "think globally and act locally". As stated by the World Bank, "Addressing the complex urban environmental problems, in order to improve urban liveability through Urban Environmental Strategies (UES), involves taking stock of the existing urban environmental problems, their comparative analysis and prioritization, setting out objectives and targets, and identification of various measures to meet these objectives.[19]

If Patrick Geddes introduced the concept of "region" in architecture planning,[20] today, the concept of "global" should be introduced to planners. The holistic theories have changed the scale of projects, and "humanity is

---

[18]It has been mentioned that a Good Anthropocene would have: System orientation, multi-scalar character, strong socio-cultural connotations-but also technology and intuitional innovations embedded, high level of popular support and knowledge in the debate on the issues, combinations of bottom-up and top-down approaches (Bennett et al., 2016).

[19]The world bank, Strategic Urban Environmental Planning available at: http://web.worl dbank.org/WBSITE/EXTERNAL/TOPICS/EXTURBANDEVELOPMENT/EXTUWM/ 0,,contentMDK:20184472~menuPK:404557~pagePK:148956~piPK:216618~theSitePK: 341511,00.html

[20]See Boardman, Philip. (1978). *The Worlds of Patrick Geddes: Biologist, Town Planner, Re-Educator, Peace-Warrior.* London; Boston: Routledge.

beginning to appreciate and attempt to manage the planet as a garden" (Weller et al., 2017).

Humankind as good Anthropocenes could manage its garden:

> "When the study of the household (ecology) and the management of the household (economics) can be merged, and when ethics can be extended to include environmental as well as human values, then we can be optimistic about the future of humankind. Accordingly, bringing together these three 'E's' is the ultimate holism and the great challenge for our future".

Eugene Odum

References

Asafu-Adjaye J., Blomqvist, L., Brand, S., Brook, B., Defries, R., Ellis, E., Foreman, C., Keith, D., Lewis, M., Lynas, M., Nordhaus, T., Pielke, J.R. R., Pritzker, R., Roy, J., Sagoff, M., Shellenberger, M., Stone, R.,Teague, P. (2015). *An Ecomodernist Manifesto.* Available at: https://static1.square space.com/static/5515d9f9e4b04d5c3198b7bb/t/552d37bbe4b07a7dd6 9fcdbb/1429026747046/An+Ecomodernist+Manifesto.pdf [Accessed 9 October 2017]

Bennett, E.M., M. Solan, R. Biggs, T. MacPhearson, A. Norstrom, P. Olsson, L. Pereira, G. D. Peterson, C. Raudsepp-Hearne, F. Beirmann, S. R. Carpenter, E. Ellis, T. Hichert, V. Galaz, M. Lahsen, B. Martin-Lopez, K. A. Nicolas, R. Preisser, G. Vince, J. Vervoort, and J. Xu. (2016). Bright Spots: Seeds of a Good Anthropocene. Frontiers in Ecology and Environment.

Corner, J. (2016). The ecological Imagination: Life in the city and Public Realm. In: Steiner, F.R., Thompson, G.F., Carbonell, A. (eds.). *Nature and Cities: The Ecological Imperative in Urban Design and Planning.* Massachusetts: Lincoln Institute of Land Policy.

Corrado, M. (2012). *Il sentiero dell'architettura porta nella foresta.* Milano: Franco Angeli.

Figueres C., Schellenhuber, H.J., Whiteman, G., Rockström, J., Hobley, A., Rahmstorf, S. (2017). Three years to safeguard our climate. *Nature 546*, p. 593-595.

Hunt, D.J. (2012). *Sette lezioni sul paesaggio.* Trans. Morabito, V. Melfi: Libria

MacDonald, D., Crabtree, J.R., Wiesinger, G., Dax, T., Stamou, N., Fleury, P., Gutierrez Lazpita, J. and Gibon, A., 2000. Agricultural abandonment in mountain areas of Europe: Environmental consequences and policy response. *Journal of Environmental Management,* 59: p. 47-69.

Marino, D. (ed.). (2016). *Agricoltura urbana e filiere corte: Un quadro della realtà italiana.* Milan: FrancoAngeli.

Muir, J. (1911). My First Summer in the Sierra. Boston: Houghton Mifflin.

Weller, R. (2015). World P-ARK. *WILD, LA+ Interdisciplinary Journal of Landscape Architecture* 1. p.10-19.

Weller, R. (2014). Stewardship Now? Reflections on Landscape Architecture's Raison d'être in the 21st Century. *Landscape Journal* 33(2). p. 1-24.

Wilson, O.E. (2006). *The creation: An Appeal to Save Life on Earth.* New York: WW. Norton and Company.

Zagari, F. (2013). *Sul paesaggio lettera aperta.* Melfi: Libria, p. 30.General References

General References

Amin, A., Thrift, N. (2002). *Cities: Reimagining the urban.* Cambridge: Polity press.

Asafu-Adjaye J., Blomqvist, L., Brand, S., Brook, B., Defries, R., Ellis, E., Foreman, C., Keith, D., Lewis, M., Lynas, M., Nordhaus, T., Pielke, J.R. R., Pritzker, R., Roy, J., Sagoff, M., Shellenberger, M., Stone, R., Teague, P. (2015). *An Ecomodernist Manifesto.* Available at: https://static1.square space.com/static/5515d9f9e4b04d5c3198b7bb/t/552d37bbe4b07a7dd6 9fcdbb/1429026747046/An+Ecomodernist+Manifesto.pdf [Accessed 9 October 2017]

Bateson, G. (1979). *Mind and Nature: A Necessary Unity (Advances in Systems Theory, Complexity, and Human Sciences).* New York: Hampton Press.

Beatley, T. (2016). *Hand book of Biophilic city planning and design.* Washington D.C.: Island press.

Benedict, M. A. and McMahon, E.T. (2006). *Green Infrastructure: Linking Landscapes and Communities.* Washington, D.C.: Island Press.

Beyer, E., Hagemann, A., Rieniets, T., Oswalt, P. (2006). *Atlas of shrinking cities.* Stuttgart: Hatje Cantz Publishers.

Brenner, N. (Ed.). (2014). *Implosions/Explosions: Towards a Study of Planetary Urbanization.* Berlin: Jovis Verlag.

Brenner, N. (2001). The limits to scale? Methodological reflections on scalar structuration. *Progress in Human Geography* 25, p. 591–614.

Brenner, N. (1998). Global cities, global states: global city formation and state 'territorial' restructuring in contemporary Europe. *Review of International Political Economy,* 5. p. 1–37.

Brosse, J. (1989). *Mythologie des arbres.* Paris: Payot.

Clément, G. (2004). *Manifeste du Triers paysage.* Paris: Éditions Sujet/Objet.

Clément, G. (1991). *Le Jardin en movement.* Paris: Pandora.

Corrado, M. (2012). *Il sentiero dell'architettura porta nella foresta.* Milano: Franco Angeli.

Czerniak, J., Hargreaves, G.(eds.) (2007). *Large parks*. New York: Princeton Architectural Press.

Di Carlo, F. (2015). Michel Corajoud and Parc Départemental du Sausset. *Journal of Landscape Architecture*, 10(3). p. 68-77.

Di Carlo, F. (2013). *Paesaggi di Calvino*. Melfi: Libria.

Elmqvist, T., Parnell,S., Fragkias, M.,Goodness, J., Guneralp, B., Scewenius, M., Sendstad, M., Macotullo, P.J., Seto, C.K., McDonald, R.I., Wilkinson, C.(eds.)(2013). *Urbanization, Biodiversity and Ecosystem Services: Challenges and Opportunities—A Global Assessment*. Dordrecht: Springer.

FAO. (2016). *Guidelines on urban and per-urban forestry*. By Salbitano F., Borelli, S., Conigliaro, M., Chen, Y. FAO Forestry Paper No.178. Rome.

Forman R.T.T. and Wu, J. (2016). Where to put the next billion people. *Macmillan Publishers Limited part of Springer Nature*, 537, p. 608-611.

Forman, R.T.T. (2015). Launching landscape ecology in America and learning from Europe. In: Barrett, G.W., Barrett T.L. Wu, J. (eds.). *History of Landscape Ecology in the United States*. New York: Springer, p. 13-30.

Forman, R.T.T. (2008). The urban region: natural systems in our place, our nourishment, our home range, our future. *Landscape Ecology* 23(3). p. 251-253.

Forman, R.T.T., Godron, M. (1986). *Landscape Ecology*. New York: John Wiley and Sons.

Ghiggi, D. (ed.) (2010). *Tree Nurseries Cultivating the Urban Jungle: Plant production Worldwide*. Baden: Lard Müller publisher.

Gouverneur, D. (2014). *Planning and design for future informal settlements*. New York: Routledge.

Grimal, P. (1974). *L'arte des Jardins*. Trans. Magi, M. Roma: Feltrinelli.

Guinaudeau, C. (1987). *Planter aujourd'hui, batir demain: le préverdissement*. Paris: Collection Mission du paysage.

Harrison, R.P. (1992). *Forests: the shadow of civilization*. Chicago: University of Chicago Press.

Hunt, D.J. (2016). *Site, Sight, Insight: Essays on Landscape Architecture*. Philadelphia: University of Pennsylvania Press.

Hunt, D.J. (2014). *Historical Ground: The role of history in contemporary landscape architecture*. 1st Edition. Abingdon: Routledge.

Hunt, D.J. (2012). *Sette lezioni sul paesaggio*. Trans. Morabito, V. Melfi: Libria

Jacobs, J. (1961). *The death and life of great American cities*. New York: Random house.

Konijnendijk, C.C., Ricard, R.M., Kenney, A., Randrup, T.B. (2006). Defining urban forestry – a comparative perspective of North America and Europe. *Urban Forestry and Urban Greening* 4, p. 93–103.

Konijnendijk C.C. (2003). A decade of urban forestry in Europe. *Forest Policy and Economics* 5, p. 173-186.

Kowarik, I., Körner, S.(eds.). (2005). *Wild urban woodlands: New perspectives for urban forestry.* Berlin: Springer.

Krajter Ostoić, S., Konijnendijk, C.C. (2015). Exploring global scientific discourses on urban forestry. *Urban Forestry & Urban Greening* 14 (2015) p. 129–138

Lawrence, H.W. (2008). *City Trees: A Historical Geography from the Renaissance through the Nineteenth Century.* Charlottesville: University of Virginia Press.

Lefbvre, L. (1970). La révolution urbaine. Paris: Gallimard

McHarg, I. (1992). *Design with nature, 25th Anniversary ed.* New York: John Wiley and Sons.

Metta, A. (2008). *Paesaggi d'autore: il Novecento in 120 progetti.* Firenze: ALINEA EDITRICE.

Miller, R.W., Hauer, R.J., Werner, L.P. (2015). *Urban Forestry: Planning and Managing Urban Greenspaces.* (Third Edition). Long Grove: Wavaland Press

Moore, C.W., Mitchell, W.J., Turnbull Jr, W. (1988). *The poetics of garden*, Trans. it. *La poetica del giardino.* Padova, Muzzio Ed., 1991.

Moore, J. W. (ed.) (2016). *Anthropocene or Capitalocene? Nature, History, and the Crisis of Capitalism.* Oakland: PM Press/Kairos.

Oelschlaeger, M. (1991). *The Idea of Wilderness: From Prehistory to the Age of Ecology.* New York: Yale University Press.

Panzini, F. (1993). *Per i piaceri del popolo. L'evoluzione del giardino pubblico in Europa dalle origini al XX secolo.* Bologna: Zanichelli Editore.

Shiva, V. (2013). *Making Peace with The Earth.* London: Pluto Press.

Shiva, V. (2005). *Earth Democracy: Justice, Sustainability and Peace.* Berkeley: North Atlantic Books.

Shiva, V. (2005). Two myths that keep the world poor. *Ode 28.* Available at: http://eprints.whiterose.ac.uk/11121/2/selmanp_natural_beauty_paper.pdf [Accessed 19August 2017]

Shiva, V. (2002). *Water Wars; Privatization, Pollution, and Profit.* California: South End Press, Cambridge Massachusetts.

Shiva, V. (2000). *Stolen Harvest: The Hijacking of the Global Food Supply.* California: South End Press, Cambridge Massachusetts.

Shiva, V. (1997). *Biopiracy: The Plunder of Nature and Knowledge.* California: South End Press, Cambridge Massachusetts.

Shiva, V. (1993). Monocultures of the Mind, Perspectives on Biodiversity and Biotechnology. London: Zed Books.

Sitte, C. *City planning According to Artistic Principals.* Trans. Collins, G.R., Collins, C.C. (1965). New York: Random house.

Steiner, F., Thompson, G.F., Carbonell, A. (2017). *Cities and Nature.* Massachusetts: Lincoln Land Institute.

Steiner, F. (2014). Frontiers in urban ecological design and planning research. *Landscape and Urban Planning* 125, p. 304–311.

Steiner, F. (2011). Landscape ecological urbanism: Origins and trajectories. *Elsevier: Landscape and urban planning*, 100. p. 333-337.

Thoren, R. (2014). Deep Roots: Foundations of Forestry in American Landscape Architecture. [online] *Scenario 04: Building the Urban Forest.* Available at: https://scenariojournal.com/article/deep-roots/ [Accessed 13 August 2017]

UN Habitat. (2013). *State of the World's Cities 2012/2013, Prosperity of Cities.* New York: Routledge.

Waldheim, C. (2016). Landscape as Urbanism: A General Theory. New Jersey: Princeton university press.

Weller, R. (2017). The City Is Not an Egg: Western Urbanization in Relation to Changing Conceptions of Nature. In: Steiner, F.R., Thompson, G.F., Carbonell, A. (eds.). *Nature and Cities: The Ecological Imperative in Urban Design and Planning.* Massachusetts: Lincoln Land Institute. p. 31-50.

Weller, R., Hands, T. (2014). Building the Global Forest. [online] *Scenario 04: Building the Urban Forest.* Available at: https://scenariojournal.com/article/building-the-global-forest/ [Accessed 15 May]

Weller, R. (2014). Stewardship Now? Reflections on Landscape Architecture's Raison d'être in the 21st Century. *Landscape Journal* 33(2). p. 1-24.

Weller, R. (2011). *Boomtown 2050: Scenarios for a Rapidly Growing City.* London: UWA Publishing.

Williams, M. (2006). *Deforesting the Earth: From Prehistory to Global Crisis.* Abridged edition. Chicago: University of Chicago Press.

Wilson E.O. (1984). *Biophilia.* Cambridge, MA: Harvard University Press

Zagari, F., Di Carlo, F. (eds.)(2016). *Il paesaggio come sfida Il Progetto.* Melfi: Libria.

Zagari, F. (2015). Moving forest. Expo Milano 2015 landscape. Melfi: Libria.

Zagari, F. (2013). *Sul paesaggio lettera aperta*. Melfi: Libria.

Zagari, F. (2009). *Giardini. Manuale di progettazione*. Roma: Mancosu Editore.

Zagari, F. (2006). *Questo è paesaggio. 48 definizioni*. Roma: Mancosu Editore.

Index

About the Author

Born in Tehran (1984), **Samaneh** is a lecturer at the Agricultural University of Iceland. She received her Ph.D. in Landscape and Environment from Sapienza - University of Rome, and her Ph.D. thesis has been elected as the winner of Europen Council of Landscape Architecture Schools (ECLAS), Annual Awards 2018 - Outstanding Student Award, 3rd cycle. She got her master with honours from Sapienza - University of Rome (2013). Before she got into Landscape studies, she completed the Bachelor of Computer Hardware Engineering (Tehran, 2002-2006). She was engaged on issues ranging from Urban renewal to Green Infrastructure and Water System management. She has conducted field research at PennDesign - Pennsylvania University (2017) and Université de Liège (2012). She has been the former teaching assistant at the Architecture Faculty of Sapienza University of Rome (2014-2019). Samaneh's research interests focus on solving Large-Scale Planning problems with particular attention to holistic approaches.

For Product Safety Concerns and Information please contact our EU
representative GPSR@taylorandfrancis.com
Taylor & Francis Verlag GmbH, Kaufingerstraße 24, 80331 München, Germany

www.ingramcontent.com/pod-product-compliance
Ingram Content Group UK Ltd.
Pitfield, Milton Keynes, MK11 3LW, UK
UKHW021821240425
457818UK00001B/19

* 9 7 8 8 7 7 0 2 2 6 5 1 6 *